Pitman Research Notes in Mathematics Series

Main Editors
H. Brezis, Université de Paris
R.G. Douglas, Texas A&M University
A. Jeffrey, University of Newcastle upon Tyne *(Founding Editor)*

T0186376

Editorial Board
H. Amann, University of Zürich
R. Aris, University of Minnesota
G.I. Barenblatt, University of Cambridge
A. Bensoussan, INRIA, France
P. Bullen, University of British Columbia
S. Donaldson, University of Oxford
R.J. Elliott, University of Alberta
R.P. Gilbert, University of Delaware
R Glowinski, University of Houston
D. Jerison, Massachusetts Institute of Technology

K. Kirchgässner, Universität Stuttgart
B. Lawson, State University of New York at Stony Brook
B. Moodie, University of Alberta
S. Mori, Kyoto University
L.E. Payne, Cornell University
G.F. Roach, University of Strathclyde
I. Stakgold, University of Delaware
W.A. Strauss, Brown University

Submission of proposals for consideration

Suggestions for publication, in the form of outlines and representative samples, are invited by the Editorial Board for assessment. Intending authors should approach one of the main editors or another member of the Editorial Board, citing the relevant AMS subject classifications. Alternatively, outlines may be sent directly to the publisher's offices. Refereeing is by members of the board and other mathematical authorities in the topic concerned, throughout the world.

Preparation of accepted manuscripts

On acceptance of a proposal, the publisher will supply full instructions for the preparation of manuscripts in a form suitable for direct photo-lithographic reproduction. Specially printed grid sheets can be provided and a contribution is offered by the publisher towards the cost of typing. Word processor output, subject to the publisher's approval, is also acceptable.

Illustrations should be prepared by the authors, ready for direct reproduction without further improvement. The use of hand-drawn symbols should be avoided wherever possible, in order to maintain maximum clarity of the text.

The publisher will be pleased to give any guidance necessary during the preparation of a typescript, and will be happy to answer any queries.

Important note

In order to avoid later retyping, intending authors are strongly urged not to begin final preparation of a typescript before receiving the publisher's guidelines. In this way it is hoped to preserve the uniform appearance of the series.

Addison Wesley Longman Ltd
Edinburgh Gate
Harlow, Essex, CM20 2JE
UK
(Telephone (0) 1279 623623)

Titles in this series. A full list is available from the publisher on request.

D Goeleven

Facultés Universitaires Notre-Dame de la Paix, Belgium

Noncoercive variational problems and related results

CRC Press
Taylor & Francis Group
Boca Raton London New York

CRC Press is an imprint of the
Taylor & Francis Group, an **informa** business

A CHAPMAN & HALL BOOK

CRC Press
Taylor & Francis Group
6000 Broken Sound Parkway NW, Suite 300
Boca Raton, FL 33487-2742

© 1996 by Taylor & Francis Group, LLC
CRC Press is an imprint of Taylor & Francis Group, an Informa business

No claim to original U.S. Government works

This book contains information obtained from authentic and highly regarded sources. Reasonable efforts have been made to
publish reliable data and information, but the author and publisher cannot assume responsibility for the validity of all materials
or the consequences of their use. The authors and publishers have attempted to trace the copyright holders of all material reproduced in this publication and apologize to copyright holders if permission to publish in this form has not been obtained. If any
copyright material has not been acknowledged please write and let us know so we may rectify in any future reprint.

Except as permitted under U.S. Copyright Law, no part of this book may be reprinted, reproduced, transmitted, or utilized in any
form by any electronic, mechanical, or other means, now known or hereafter invented, including photocopying, microfilming,
and recording, or in any information storage or retrieval system, without written permission from the publishers.

For permission to photocopy or use material electronically from this work, please access www.copyright.com (http://www.copy-
right.com/) or contact the Copyright Clearance Center, Inc. (CCC), 222 Rosewood Drive, Danvers, MA 01923, 978-750-8400.
CCC is a not-for-profit organization that provides licenses and registration for a variety of users. For organizations that have been
granted a photocopy license by the CCC, a separate system of payment has been arranged.

Trademark Notice: Product or corporate names may be trademarks or registered trademarks, and are used only for identifica-
tion and explanation without intent to infringe.

Visit the Taylor & Francis Web site at
http://www.taylorandfrancis.com

and the CRC Press Web site at
http://www.crcpress.com

Contents

PREFACE

In a Remarkably short time, the field of noncoercive problems has seen considerable development in Mathematics and Mechanics.

This tract deals with the study of variational problems which can be formulated as the minimization of some possibly nonconvex and noncoercive functionals on a possibly nonconvex set. Some additional results and possible prospects concerning the field of nonvariational problems are also considered. A recession approach relying on the asymptotic behavior of the data involved in the problem is combined with a Tychonov regularization. Here, the recession approach has been reviewed and a quite general and constructive theory is given. The flexibility of the approach is then illustrated by its application to several problems in Mathematics and Mechanics.

The main object of this paper is to present a general mathematical theory applicable to the study of noncoercive unilateral problems arising in Nonsmooth Mechanics. Particularly, we direct our efforts to the study of problems which can be formulated by means of variational inequalities (approach of G. Duvaut and J.L. Lions [55] and J.-J. Moreau [102]) and hemivariational inequalities (approach of P.D. Panagiotopoulos [112]). However, other approaches and problems are considered too since the scope of this work is also to recover or generalize some known results and to connect several theories.

It goes without saying that this paper has considerably been influenced by the works of numerous specialists. For instance, the works of C. Baiocchi, G. Buttazzo, H. Brézis, G. Fichera, F. Gastaldi, A. Haraux, P. Hess, L. Nirenberg, M. Schatzman, F. Tomarelli, etc. in originating and developing methods for the treatment of noncoercive problems have considerably inspired the present work.

These original works have here been reviewed with a thought of unification and the consideration of recent ideas and developments among which the ones due to S. Adly, D.D. Ang, H. Attouch, A. Auslender, D. Goeleven, J. Mawhin, Z. Naniewicz, P.D. Panagiotopoulos, K. Schmitt, M. Théra, Vy K. Le, etc.

The original idea of the recession approach goes back to G. Fichera [61] whose work deals with the study of semicoercive variational inequalities. As a consequence of the contributions of G. Fichera, the study of noncoercive variational problems, variational inequalities and hemivariational inequalities has emerged as an important branch of Applied Mathematics and Mechanics.

This work has been partly carried out when I was visiting the University of Montpellier II and the University of Utah. I would like to express my gratitude to Professor H. Attouch (Montpellier) and Professor K. Schmitt (Utah) for their hospitalities. I also wish to acknowledge the helpful Remarks and suggestions received from Dr. S. Adly (Limoges), Professor G. Buttazzo (Pisa), Professor G. Gagneux (Pau), Professor V.H. Nguyen (Namur), Professor D. Motreanu (Iasi), Professor J.-P. Penot (Pau), Professor K. Schmitt (Utah), Professor M. Théra (Limoges) and Dr. Vy K. Le (Utah). I wish to express my sincere thanks to Professor P.D. Panagiotopoulos who suggested me for research the new and fertile direction of hemivariational inequalities in the field of inequality problems. I would like to thank my editors in Longman for their remarkable assistance. Finally, I wish to thank the F.N.R.S. for financial support.

I must apologize to those whose work has been inadvertently neglected in the list of references given in this paper. I shall welcome all comments and corrections from readers.

1 INTRODUCTION

In recent years, many engineers and mathematicians have directed their attention towards noncoercive problems. The reason is that various engineering models lead, generally, to a noncoercive problem. The lack of coercivity may be due to boundary conditions which are insufficiently blocking up or to the presence of a destabilizing term depending on a parameter as it is the case in the unilateral buckling in elasticity.

One of the first important noncoercive problem is maybe the one given by the elliptic boundary value problem at resonance which has been investigated by E.M. Landesman and A.C. Lazer [91].

The results of E.M. Landesman and A.C. Lazer have been extended in various directions by a quite big number of authors (see L. Boccardo, P. Drabek and M. Kučera [25], P. Drabek [54], H. Brézis and L. Nirenberg [33], S. Fučik [63] and J. Mawhin [97] where many references may be found). A more general theory concerning the study of noncoercive and nonlinear partial differential equations has been set up by H. Brézis and L. Nirenberg [33]. In their paper, H. Brézis and L. Nirenberg introduce the recession function associated to a general nonlinear operator.

Several related results have then been improved for the specific study of equations governed by a noncoercive maximal monotone operator or an accretive operator. See for instance H. Brézis and A. Haraux [34], A. Haraux [80], B.D. Calvert and C.P. Gupta [39] and the recent theory of H. Attouch, Z. Chbani and A. Moudafi [14].

The recession functions are well known by the mathematicians working in optimization and convex analysis, see for instance J.P. Dedieu [52], D.T. Luc [94], [95], D.T. Luc and J.P. Penot [96] and R.T. Rockafellar [124]. However, the importance of this tool in nonlinear analysis and mechanics has been pointed out by C. Baiocchi, G. Buttazzo, F. Gastaldi and F. Tomarelli [21]. In their paper, the problem of minimizing a possibly nonconvex and noncoercive functional is studied. See also the other publications [20], [38] and [64] of the same school.

The limit load problem in plasticity is also an important problem requiring the concept of recession function, see G. Bouchitte and P. Suquet [26] and R. Temam [130].

The original idea of the recession approach (by a recession approach, we mean an approach relying on the asymptotic behavior of the sets, functions and operators which are involved in a problem) seems to go back to G. Fichera [61] whose work deals with the study of linear semicoercive variational inequalities. See also the paper of M. Schatzman [125] which is in some sense the generalization of the work of G. Fichera to noncoercive and nonlinear monotone variational inequalities. Many important applications in mechanics of the results of G. Fichera and M. Schatzman can be found in the book of P.D. Panagiotopoulos [112].

For further results related to a recession approach concerning noncoercive variational inequalities, we refer to D.D. Ang, K. Schmitt and L.K. Vy [6]-[8], G. Duvaut and J.L. Lions [55], D. Goeleven [67], P. Hess [82], J.L. Lions and G. Stampacchia [93], P. Shi and M. Shillor [126] and the references cited therein. For a general recession theory applicable to a large class of nonlinear variational inequalities involving pseudomonotone operators, we refer to the paper of S. Adly, D. Goeleven and M. Théra [3] and the one of F. Tomarelli [132].

Many problems in mechanics which are connected to nonsmooth and nonconvex functionals can be formulated as a hemivariational inequality. That is a general formulation (including variational inequalities) introduced by P.D. Panagiotopoulos [110], [111]. Most of these problems are also noncoercive and have been studied for the first time by P.D. Panagiotopoulos [117], [118]. Some of these results have been reviewed and improved by using the recession approach (see D. Goeleven and M. Théra [79]).

More recently, the recession approach for variational problems has been reviewed by A. Auslender [18] and by L.K. Vy and K. Schmitt [135]. The work of A. Auslender is a generalization of the one of C. Baiocchi, G. Buttazzo, F. Gastaldi and F. Tomarelli [21]. The paper of L.K. Vy and K. Schmitt concerns the minimization of noncoercive homogeneous functionals on manifolds.

In this paper, we present a general recession method for the problem of minimizing a possibly nonconvex and noncoercive functional on a possibly nonconvex set. The abstract recession approach is a variant of the known ones cited above but is built in such a way that it enables us to discuss simply many different kinds of variational problems. We will show the flexibility of the approach by applying our theoretical results to many different mathematical and mechanical problems. Some known results will be recovered and generalized by following a unique method (Sections 4.1-4.3, 4.5, 4.7-4.8, 4.12, 4.13) and several new results for problems which have not yet often been studied in the literature will be obtained (Sections 4.4, 4.6, 4.9-4.11).

The approach presented could also be extended so as to be applicable to variational inequalities and hemivariational inequalities which also hold in nonvariational situations. It is out of the scope of this work to consider all the possible extensions of the recession analysis to nonvariational problems, that are problems which cannot be reduced to the minimization of some functionals. Such considerations could be the object of another tract. However, in Section 3.3, we will investigate a general class of nonvariational hemivariational inequalities invoking a nonconvex set of constraints. This last class of problems is sufficiently general to reflect how the recession approach can be used to treat nonvariational problems. For more details concerning the study of various general classes of variational and hemivariational inequalities, we refer to S. Adly, D. Goeleven and M. Théra [3], D. Goeleven and P.D. Panagiotopoulos [76] and D. Goeleven and M. Théra [79].

Definitions and general properties of the recession functions, weak-coercivness conditions and other preliminaries are given in Section 2.

In Section 3.1, we consider an abstract minimization problem in a general framework and we combine a Tychonov regularization with a general recession approach

3

in order to present a constructive theory. In Section 3.2, the results of Section 3.1 are reviewed when we replace the Tychonov regularization by a general viscosity approach. A new information is then given by the fact that the solution obtained satisfies a viscosity selection principle [12]. It is important to notice that our assumptions are such that the regularization invoked by the recession process lead to a solution of the original problem. However, there exist several problems for which a relaxation phenomena may occur. Such problems are not considered in this paper and we refer the interested reader to the work of G. Buttazzo [37], a recent paper of H. Attouch [12] and references cited therein.

As stated above, Section 3.3 concerns the extension of the recession analysis for the study of nonvariational problems invoking possibly nonconvex sets of constraints. It is based on recent results due to Z. Naniewicz [106] and D. Goeleven [71].

Section 4 deals with the applications of the abstract recession method. We will show that a unique methodology can be used in order to study many different problems. In Section 4.1, we consider an elliptic problem with unilateral boundary conditions. Section 4.2 deals with the study of a model proposed by M. Giaquinta and E. Giusti [65] for the description of masonry-like problem. In Section 4.3, we recover the theory of C. Baiocchi, G. Buttazzo, F. Gastaldi and F. Tomarelli [21]. In Section 4.4, we refine our recession analysis when a separability property holds for the 'energy' functional so that we are able to exploit some orthogonal decomposition of the space.

The three-dimensional nonlinear elasticity model proposed by J.M. Ball [22] has been intensively studied in the literature. See [1], [21], [22], [43]-[45], [50], [62] and the references cited therein. However, most of the mathematical results are only applicable to bilateral and coercive problems. Section 4.5 deals with the study of noncoercive unilateral problems involving locking constraints which have been introduced and studied for the first time by P.G. Ciarlet and J. Necas [44]. A similar problem formulated by using the theory of P.D. Panagiotopoulos is again considered in Section 4.10. In Section 4.6, we show how weak convergence methods of L.C. Evans [57] can be used in order to check a compactness condition used in the recession analysis. In Section 4.7, we recover the theory of K.L. Vy and K. Schmitt [135]. Section 4.8 deals with the study of some nonlinear partial differential equations.

In Section 4.9, 4.10, 4.11 and 4.12, we use the recession analysis in order to prove new results in the theory of variational inequalities and hemivariational inequalities.

Section 4.13 ends this paper with a very simple linear ill-posed system in order to show the properties of the solutions obtained by the recession method.

A list of the main notations used throughout the text is given at the end of this paper.

2 PRELIMINARIES

2.1 Recession tools

If C is a convex set of $I\!\!R^N (N \in I\!\!N \backslash \{0\})$ then its behavior at infinity can be described in terms of what is called the recession directions, i.e. the directions which recede from C. The cone of recession directions is called the recession cone of C and usually denoted by C_∞, that is

$$C_\infty := \{y \in I\!\!R^N \mid x + \mu y \in C, \forall \mu \geq 0, \forall x \in C\}.$$

For examples of recession cones of convex sets in $I\!\!R^2$, for

$$A = \{(x_1, x_2) \in I\!\!R^2 \mid x_1 > 0, \ x_1 x_2 \geq 1\},$$
$$B = \{(x_1, x_2) \in I\!\!R^2 \mid x_2 \geq x_1^2 + 1\},$$

one has

$$A_\infty = \{(x_1, x_2) \in I\!\!R^2 \mid x_1 \geq 0, \ x_2 \geq 0\},$$

and

$$B_\infty = \{(x_1, x_2) \in I\!\!R^2 \mid x_1 = 0, \ x_2 \geq 0\}$$

If $g : I\!\!R^N \to I\!\!R \cup \{+\infty\}$ is a proper and convex function, then its behavior at infinity is described in terms of the asymptotic behavior of its epigraph. The recession function of g is denoted by g_∞ and is defined as the function whose epigraph is the recession cone of the epigraph of g, i.e.

$$\text{epi}(g_\infty) = (\text{epi}(g))_\infty.$$

For examples of recession functions of convex functions on $I\!\!R^2$, for

$$f(x_1, x_2) = x_1^2 + x_2^2,$$
$$h(x_1, x_2) = ln(e^{x_1} + e^{x_2}),$$

one has

$$f_\infty(y_1, y_2) = \begin{cases} 0 & \text{if } (y_1, y_2) = (0,0) \\ +\infty & \text{otherwise}, \end{cases}$$

and

$$h_\infty(y_1, y_2) = \max\{y_1, y_2\}.$$

These concepts are intensively studied in the book of R.T. Rockafellar [124]. See also the book of N. Bourbaki [28] for arbitrary topological vector spaces.

By using a topological approach, these notions of recession functionals and recession cones have been generalized by C. Baiocchi, G. Buttazzo, F. Gastaldi and F. Tomarelli [21] in order to cover the cases of nonconvex functionals and nonconvex sets.

Let (X, τ) be a real Hausdorff topological space endowed with a topology τ and let $G : X \rightarrow \mathbb{R} \cup \{+\infty\}$ be any functional. We call τ-recession function of G [21] the function (we restrict our study to the sequential topological concept)

$$\tau - G_\infty(x) : = \inf\{\liminf G(t_n x_n)/t_n \mid t_n \rightarrow +\infty, x_n \xrightarrow{\tau} x\}.$$

We will use the notations $w - G_\infty$ and G_∞ whenever τ denotes respectively the weak $\sigma(X, X')$ ($\rightharpoonup$) and the strong ($\rightarrow$) topology in a real Banach space.

The following Proposition lists some basic properties of the recession funtional.

Proposition 2.1.1. ([21]; Baiocchi-Buttazzo-Gastaldi-Tomarelli)
Let $G, H : X \rightarrow \mathbb{R} \cup \{+\infty\}$ be functionals defined on X. Then

(1) $\tau - G_\infty$ is τ-lower semicontinuous and positively homogeneous of order 1;

(2) $\tau - (G + H)_\infty \geq \tau - G_\infty + \tau - H_\infty$;

(3) if H is positively homogeneous of degree 1 and τ-continuous, then

$$\tau - (G + H)_\infty = \tau - G_\infty + H;$$

(4) if G is nonnegative, positively homogeneous of degree greater than 1 and τ-lower semicontinuous, then

$$
\tau - G_\infty(x) \;=\; \begin{cases} +\infty & \text{if } G(x) \neq 0 \\[2ex] 0 & \text{if } G(x) = 0; \end{cases}
$$

(5) if

$$
\inf\{G(u) \mid u \in X\} > -\infty
$$

then

$$
\tau - G_\infty(u) \geq 0, \forall\, u \in X.
$$

To illustrate the above concept, let us consider the following example

$$
h(x) = \tfrac{1}{2}\langle Ax, x\rangle + \langle a, x\rangle + \alpha,
$$

where $A : X \to X$ is a nonnegative and symmetric bounded linear operator, X is a real Hilbert space, $a \in X$ and $\alpha \in \mathbb{R}$. Using properties (3) and (4) of Proposition 2.1.1, we obtain easily

$$
h_\infty(x) = \begin{cases} \langle a, x\rangle & \text{if } x \in Ker\,A \\ +\infty & \text{if } x \notin Ker\,A \end{cases}
$$

If in addition G is convex then we have some properties more. We list some of them in the following Proposition.

Proposition 2.1.2. Let $G : X \to \mathbb{R} \cup \{+\infty\}$ be a proper, convex and τ-lower semicontinuous function. Then

(1) G_∞ is proper convex and τ-lower semicontinuous. Moreover

$$
\tau - G_\infty(x) \;=\; G_\infty(x) \;=\; \lim_{\lambda \to +\infty} [G(x_o + \lambda x) - G(x_o)]/\lambda
$$

where x_o is an arbitrary element of $\mathrm{dom}(G)$;

8

(2)

$$G(u + v) \leq G(u) + G_\infty(v), \forall u, v \in X;$$

For the proof of the properties listed above, we refer to the article of C. Baiocchi, G. Buttazzo, F. Gastaldi and F. Tomarelli [21] and the book of G. Buttazzo [37].

Now, let us define the concept of τ-recession cone of a set. Let K be a subset of X. The τ-recession cone of K is the τ-closed cone (not convex in general) defined by

$$\tau - K_\infty := \{x \in X \mid \exists \{t_n\}, \exists \{x_n\} \text{ such that } t_n \to +\infty,$$

$$x_n \xrightarrow{\tau} x \text{ and } t_n x_n \in K, \forall n \in \mathbb{N}\}.$$

The following Proposition lists the main properties of the recession cone.

Proposition 2.1.3. ([21]; Baiocchi-Buttazzo-Gastaldi-Tomarelli) Let K be a nonempty subset of X.

(1) If K is bounded then $\tau - K_\infty = \{0\}$;

(2) we have

$$\tau - K_\infty = \tau - (K \backslash B)_\infty = \tau - (K \cup B)_\infty,$$

for all nonempty bounded subset B of X;

(3)

$$\tau - K_\infty = \text{dom}(\{\psi_K\}_\infty),$$

where ψ_K denotes the indicator function of K;

(4)

$$\tau - K_\infty = \cap_{\mu > 0} cl_\tau [\cup_{\lambda > \mu} \frac{1}{\lambda} (K - x_0)]$$

where x_o is any point of X and cl_τ denotes the topological closure with respect to τ;

9

(5) if K is a τ-closed and convex subset of X then K_∞ is convex,

$$\tau - K_\infty \;=\; K_\infty$$

and

$$K_\infty \;=\; \cap_{\mu>0}(\frac{1}{\mu}(K - x_0))$$

where x_o is any element of K;

(6) assume that $G : X \to \mathbb{R} \cup \{+\infty\}$ is a proper, convex and lower semicontinuous function. If K is a nonempty and closed set, then

$$(G + \psi_K)_\infty \;=\; G_\infty + (\psi_K)_\infty \;=\; G_\infty + \psi_{K_\infty};$$

Let $A : X \to X'$ be a given operator. The concepts introduced above are related to the concept of recession function in the sense of H. Brézis and L. Nirenberg [33], i.e.

$$
\begin{aligned}
J_A(u) \;&:=\; \liminf_{\substack{t\to+\infty \\ v\to u}}\langle A(tv), v\rangle \\
\;&=\; \inf\{\liminf\langle A(t_n v_n), v_n\rangle \mid t_n \to +\infty, v_n \to u\}.
\end{aligned}
$$

If we set $G(x) = \langle Ax, x\rangle$, then we have

$$J_A(u) \;=\; G_\infty(u).$$

More generally, let $u_0 \in X$ be given. We introduce the recession function of A with respect to u_0 by the formula [71]:

$$r_{u_0,A}(u) \;:=\; \liminf_{\substack{t\to+\infty \\ v\to u}}\langle A(tv), tv - u_0\rangle/t.$$

If we set $H(x) := \langle Ax, x - u_0\rangle$, then

$$r_{u_0,A}(u) \;=\; H_\infty(u).$$

Moreover, it is clear that

$$r_{0,A}(u) = J_A(u).$$

We will see later in Section 3.3 that these last tools can be used to develop a recession approach for the study of nonvariational problems. Two important results are considered in the following two Propositions.

Proposition 2.1.4. ([33]; Brézis-Nirenberg) If B is monotone and sublinear, i.e. $\| Bv \| / \| v \| \to 0$ as $\| v \| \to +\infty$, then

$$J_B(u) = \sigma_{R(B)}(u)$$

where $\sigma_{R(B)}(u)$ is the support function of $R(B)$. In particular, J_B is convex. In addition, if $R(B)$ is bounded, then J_B is continuous.

Proposition 2.1.5. ([33]; Brézis-Nirenberg) Suppose that X is a real Hilbert space. Let $B : X \to X'$ be a nonlinear operator and let N be a finite dimensional subspace of X. The following relations are equivalent

$$\sigma_{R(B)}(v) > \langle f, v \rangle, \forall v \in N \backslash \{0\}, \tag{1}$$

$$f \in \text{int}\{N^{\perp} + conv\ R(B)\}. \tag{2}$$

Remarks 2.1.1.

i) We have

$$\begin{aligned}\sigma_{R(B)}(u) &= \sup\{\langle v, u \rangle \mid v \in R(B)\} \\ &= \sup\{\langle v, u \rangle \mid v \in \overline{conv\ R(B)}\}.\end{aligned}$$

ii) We always have

$$J_B(u) \leq \sigma_{R(B)}(u).$$

11

We close this Section by mentioning a further recession tool which has been introduced by P.L. Lions [92] and more recently studied by H. Attouch, Z. Chbani and A. Moudafi [14]. Given a set-valued operator $T : X \to 2^{X'}$, we define the recession set $\tau - G_\infty(T) \subset X \times X'$ by saying that $(u, v) \in \tau - G_\infty(T)$ if and only if there exist sequences $\{\lambda_n; n \in I\!N\}$ and $\{(u_n, v_n); n \in I\!N\}$ such that $\lambda_n \to 0, (u_n, v_n) \in$ graph$(T), \lambda_n u_n \xrightarrow{\tau} u$ and $v_n \xrightarrow{\tau} v$. If the operator $T : X \to 2^{X'}$ is maximal monotone, then a result of P.L. Lions [92] and H. Attouch, Z. Chbani and A. Moudafi [14] permits us to define a maximal monotone operator T_∞ such that

$$G_\infty(T) = \text{graph}(T_\infty) = \partial \sigma_{R(T)}.$$

Proposition 2.1.6. ([14]; Attouch-Chbani-Moudafi) The range of T_∞ is closed and

$$R(T_\infty) = T_\infty(0) = \overline{R(T)}$$

We say that the operator $T : X \to 2^{X'}$ satisfies the Brézis-Haraux condition [34] if $\forall f \in R(T), \forall y \in D(T)$:

$$\sup\{\langle h - f, y - z\rangle; (z, h) \in \text{graph}(T)\} < +\infty.$$

Remark 2.1.2. Conditions guaranteeing that a given operator satisfies the Brézis-Haraux condition are given in [34] and [80]. For instance, if T is the subdifferential of a convex function or if T is the resolvent of a maximal monotone operator, then the Brézis-Haraux condition is satisfied.

Proposition 2.1.7. ([14]; Attouch-Chbani-Moudafi) Let $A, B : X \to 2^{X'}$ be two maximal monotone operators. Suppose that

(i) A and B satisfy the Brézis-Haraux condition;

(ii) $\overline{A + B}$ is maximal monotone.

12

Then

$$\sigma_{R(A+B)} = \sigma_{R(A)} + \sigma_{R(B)}.$$

Proposition 2.1.8. ([14]; Attouch-Chbani-Moudafi) Let $T = \partial G$ be the subdifferential of a convex, lower semicontinuous, proper function $G : X \to \mathbb{R} \cup \{+\infty\}$. Then

$$T_\infty = \partial G_\infty$$

and

$$\sigma_{R(T)} = G_\infty.$$

Moreover, if $0 \in D(T)$ then

$$J_T = \sigma_{R(T)}.$$

The recession operator T_∞ can be used to develop a unified recession approach applicable to the study of possibly nonvariational problems invoking maximal monotone operators. Note that in the previous Proposition, the recession mapping J_T is defined for a set-valued operator [14]. For more details we refer the interested reader to the article of H. Attouch, Z. Chbani and A. Moudafi [14].

2.2 Weak-coercivity

Let X be a real Banach space such that either X is reflexive or $X = V'$ with V separable and τ is the weak* topology of X. Let $G : X \to \mathbb{R} \cup \{+\infty\}$ be any functional. We say that G is coercive if

$$G(u) \to +\infty \text{ as } \| u \| \to +\infty.$$

For some classes of functions, the coercivity entails some strong properties on the corresponding recession function.

Proposition 2.2.1. ([26]; Bouchitte-Suquet) If $G : X \to I\!\!R \cup \{+\infty\}$ is proper, convex, τ-lower semicontinuous and coercive. Then

(1) G_∞ is coercive

(2) $G_\infty(u) > 0, \forall u \in X\backslash\{0\}$.

If in addition G can be written as follows

$$G(u) = F(u) - \lambda L(u),$$

where F is a proper, convex, coercive and τ-lower semicontinuous functional, L is a linear and τ-continuous form and λ is a positive real number then the following relations (3) and (4) are equivalent:

$$G \text{ is coercive} \tag{3}$$

$$\lambda < \lambda_c := \min\{F_\infty(u) \mid u \in X, L(u) = 1\}. \tag{4}$$

Let X be a real Hilbert space. Let $X_1 \subset X$ be a closed vector subspace and $X_o = X_1^\perp$. For $u \in X$, let us write $u = u_o + u_1$ with $u_o \in X_o$ and $u_1 \in X_1$. We shall say that the function G is X_1 -coercive (in the sense of J. Mawhin [98]) if there exists $\alpha : I\!\!R_+ \to I\!\!R$ such that $\alpha(t) \to +\infty$ and

$$G(u) \geq \alpha(\| u_1 \|) \| u_1 \|, \forall u \in X. \tag{5}$$

For instance, let $A : X \to X'$ be a bounded linear and symmetric operator. We say that A is semicoercive on X if

$$\langle Au, u \rangle \geq \alpha \| u \|^2, \forall u \in Ker(A)^\perp.$$

In this case, the functional $G : X \to I\!\!R$ defined by

$$G(x) = \langle Ax, x \rangle$$

is $Ker(A)^\perp$-coercive.

14

Let $(H, |\,.\,|)$ be a real Hilbert space such that X is compactly embedded in H. We say that $G : X \to \mathbb{R} \cup \{+\infty\}$ is X-coercive with respect to H, if there exist $\alpha_1 \geq 0, \alpha_2 > 0$ such that

$$G(u) + \alpha_1 \mid u \mid^2 \geq \alpha_2 \| u \|^2, \forall u \in X.$$

Suppose that $G(x) = \langle Ax, x \rangle$, where A is a bounded linear symmetric and nonnegative operator. If G is X-coercive with respect to H, then

$$\dim\{Ker(A)\} < +\infty$$

and

$$A \text{ is semicoercive on } X.$$

For a proof of this result, we refer to B. Heron and M. Sermange [81] or D. Goeleven [68].

2.3 Lower semicontinuous functionals and functionals of class (Q), (Q⁺) and (A)

In this Section, we introduce several classes of functionals which will be considered later in this work.

Let (X, τ) be a topological space and let $\psi : X \to \mathbb{R} \cup \{+\infty\}$ be any functional.

Definition 2.3.1. We say that ψ is τ-lower semicontinuous if for every $t \in \mathbb{R}$ the set

$$\{u \in X : \psi(u) \leq t\}$$

is τ-closed in X.

Some basic properties are recalled in the following Proposition (see for instance [37] or [122]).

15

Proposition 2.3.1.

i) ψ is τ-lower semicontinuous if and only if the epigraph of ψ, i.e.

$$\text{epi}(\psi) := \{(x,t) \in X \times \mathbb{R} \cup \{+\infty\} \mid \psi(x) \leq t\}$$

is τ-closed in $X \times \mathbb{R} \cup \{+\infty\}$.

ii) A subset C of X is τ-closed if and only if the indicator function of C is τ-lower semicontinuous.

iii) If F and G are τ-lower semicontinuous functionals from X into $\mathbb{R} \cup \{+\infty\}$, then $F + G$ is τ-lower semicontinuous.

Remark 2.3.1. For more details concerning semicontinuity in the calculus of variation, we refer to the work of G. Buttazzo [37] and B. Dacorogna [50].

The following two classes of functionals have been introduced in a more general form by K. Schmitt and L.K. Vy [135] and will also be involved later in this work. Let $(X, \| . \|)$ be a real Banach space.

Definition 2.3.2. We say that a functional ψ has property (Q) on C whenever the following holds: If $\{u_n; n \in \mathbb{N}\} \subset C$ is any sequence in C satisfying

$$\| u_n \| \to +\infty,$$

$$w_n := u_n / \| u_n \| \rightharpoonup 0,$$

$$\limsup \psi(u_n) / \| u_n \|^2 \leq 0,$$

then there exists $v_o \in C$ such that

$$\limsup \psi(u_n) > \psi(v_o).$$

Definition 2.3.3. We say that a functional ψ has property (Q^+) on C whenever the following holds: There exists a constant $p > 1$ such that: If $\{u_n; n \in I\!N\} \subset C$ is any sequence in C satisfying

$$\| u_n \| \to +\infty,$$
$$w_n := u_n/ \| u_n \| \rightharpoonup w,$$
$$\| u_n \| \leq \| u_n - \lambda w \|, \forall \lambda \geq 1,$$
$$\limsup \psi(u_n)/ \| u_n \|^p \leq 0,$$

then there exists $v_o \in C$ such that

$$\limsup \psi(u_n) > \psi(v_o).$$

At last, we close this Section with a class of functionals whose idea goes back to A. Auslender [18].

Definition 2.3.4. We say that a functional ψ has property (A) on C whenever the following holds: If $\{u_n; n \in I\!N\} \subset C$ is a sequence such that

$$\| u_n \| \to +\infty,$$
$$w_n := u_n/ \| u_n \| \overset{\tau}{\to} w,$$
$$\limsup \psi(u_n)/ \| u_n \|^p \leq 0, \forall p > 0,$$

then there exists a subsequence (again denoted by $\{u_n, n \in I\!N\}$) and sequences $\{z_n, n \in I\!N\} \subset X$ and $\{\mu_n, n \in I\!N\} \subset (0, \| u_n \|]$ such that

$$z_n \to z,$$
$$\| z - w \| < 1,$$
$$u_n - \mu_n z_n \in C$$

and

$$\psi(u_n - \mu_n z_n) \leq \psi(u_n).$$

17

Remark 2.3.2. Various examples of functionals satisfying properties (Q) and (Q^+) will be considered in Section 4.7. ii) Set $C = I\!R$ and let $\psi : I\!R \to I\!R$ be the function defined by

$$\psi(x) = \begin{cases} e^{-x^2} & \text{if } x \in (-\infty, -1) \cup (+1, +\infty) \\ \mid (1 + e^{-1})x \mid -1 & \text{if } x \in [-1, +1]. \end{cases}$$

We can see that ψ has property (A) on $I\!R$. Indeed, let $u_n \in I\!R$ be a sequence such that $u_n/ \mid u_n \mid \to w$ and $\mid u_n \mid \to +\infty$. We choose $z_n = \frac{2}{4}w$ and $\mu_n = \frac{4}{5}(u_n - Sign(w)\alpha)/w$ with $\alpha > 0$ chosen so that $\psi(\alpha) < 0$. If $u_n \to +\infty$ then $w = +1$, $\mu_n = \frac{4}{5}(u_n - \alpha)$ so that for n great enough, we have $0 < \mu_n \leq u_n = \mid u_n \mid$. If $u_n \to -\infty$ then $w = -1, \mu_n = \frac{4}{5}(-\alpha - u_n)$ and thus for n great enough, we have also $0 < \mu_n \leq \mid u_n \mid$. It is clear that $z_n \to z = \frac{2}{4}w$ and $\| w - z \| < 1$. Finally, we have $\psi(u_n - \mu_n z_n) = \psi(Sign(w)\alpha) < 0 \leq \psi(u_n)$ for n great enough.

2.4 The recession set

Let $\{\Delta_n; n \in I\!N\}$ be a sequence of nonempty subsets of X. We denote by $R(\Delta_n)$ the set given by

$$R(\Delta_n) := \{w \in X \mid \exists u_n \in \Delta_n, \| u_n \| \to +\infty$$
$$\text{and } w_n := u_n/ \| u_n \| \xrightarrow{\tau} w\}.$$

It is worthwhile to Remark that the notation "$R(\Delta_n)$" is misuse in the sense that the set $R(\Delta_n)$ does not depend of n but of the whole sequence $\{\Delta_n; n \in I\!N\}$. The set $R(\Delta_n)$ is called the recession set associated to the sequence $\{\Delta_n; n \in I\!N\}$ and it is clear that

$$R(\Delta_n) \subset r(\Delta_n)$$

where

$$r(\Delta_n) := \{w \in X \mid \exists u_n \in \Delta_n, \exists t_n \to +\infty$$
$$\text{such that } w_n := u_n/t_n \xrightarrow{\tau} w\}.$$

18

Remark 2.4.1.

i) If $\Delta_n \equiv \Delta, \forall n \in \mathbb{N}$, then $r(\Delta_n) = \tau - \Delta_\infty$.

ii) Let $T : X \to 2^{X'}$ be a maximal monotone operator. If Δ_n is defined by saying that $u_n \in \Delta_n$ if and only if $(u_n, v_n) \in \text{graph}(T)$ for some $v_n \in X$ such that $v_n \xrightarrow{\tau} v$, then $r(\Delta_n) = \partial \sigma_{R(T)}$ (see Section 2.1).

iii) If X is a reflexive Banach space or $X = V'$ with V separable and τ is the weak* topology of X, then $R(\Delta_n) = \emptyset$ if and only if there exists $K > 0$ such that $\| v \| \leq K, \forall v \in \Delta_n, \forall n \in \mathbb{N}$.

The abstract theory which will be presented in this work is based on the properties of a recession set defined by means of a viscosity method. It is important to Remark that we can also use other standard methods. It depends on the choice of the sequence $\{\Delta_n; n \in \mathbb{N}\}$. For instance, let us consider the variational problem:

$$\min\{\psi(x) \mid x \in C\}.$$

Then we can define the sets Δ_n so as to deal with either the Galerkin method or the level set method.

a) Galerkin method.

We choose

$$
\begin{aligned}
\Delta_n \quad &:= \quad \text{argmin}\{\psi(x) \mid x \in C_n\} \\
&= \quad \{z \in C_n \mid \psi(z) \leq \psi(y), \forall y \in C_n\}
\end{aligned}
$$

where the sequence $\{C_n; n \in \mathbb{N}\}$ defines a Galerkin aproximation of the set C [134].

19

b) Level set method.

We choose

$$\Delta_n \equiv \Delta := \{x \in C \mid \psi(x) \leq \psi(x_o)\}$$

with $x_o \in C$.

For further discussions concerning the recession set, we refer to the article of D.T. Luc and J.P. Penot [96].

The same formulation can be used for the study of more general variational inequalities and hemivariational inequalities. For instance, suppose that C is a closed convex set containing 0. Let us consider the variational inequality

$$u \in C : \langle Au - f, v - u \rangle \geq 0, \forall v \in C.$$

Then, we choose

$$\Delta_n \equiv \Delta := \{x \in C \mid \langle Ax, x \rangle \leq \langle f, x \rangle\}.$$

See for instance the paper of F. Tomarelli [132] and the one of S. Adly, D. Goeleven and M. Théra [3]. These aspects will also be considered with more details in Section 3.3.

3 ABSTRACT RECESSION ANALYSIS

3.1 Recession analysis involving a Tychonov Regularization

Suppose that the assumptions (h) described below are satisfied.

(h_1) X is a Banach space such that either X is reflexive or $X = V'$ with V separable and τ is the weak* topology of X.

(h_2) C is a nonempty τ-closed subset of X;

(h_3) $\psi : X \rightarrow I\!R \cup \{+\infty\}; x \rightarrow \psi(x)$ is a τ-lower semicontinuous function;

(h_4) $\psi(x) \geq -\alpha_1 \parallel x \parallel^\beta -\alpha_2$ ($\alpha_1 \geq 0, \alpha_2 \in I\!R, 0 < \beta < 2$);

(h_5) $\text{dom}\{\psi\} \cap C \neq \emptyset$.

Banach spaces considered in (h_1) are endowed with a topology τ such that closed balls are sequentially τ-compact. We have in mind the weak topology in a reflexive Banach space and the weak* topology in the space of measure with bounded total variation. Typical examples of this situation considered in this work are: a) $X = W^{1,p}(\Omega)$ (with $p > 1, \Omega \subset I\!R^N$), τ the weak topology on X, b) $X = BV(\Omega) = \{u \in L^1(\Omega), \frac{\partial u}{\partial x_i} \in M^1(\Omega), i = 1, ..., N\}$ (where $M^1(\Omega)$ stands for the space of bounded measures on Ω), τ is the weak* topology on X, c) $X = BD(\Omega) = \{u \in L^1(\Omega; I\!R^N), \varepsilon_{ij}(u) \in M^1(\Omega), i, j = 1, ..., N\}$, τ is the weak* topology on X.

We will now consider the recession approach with a standard Tychonov regularization [133].

Let $\{\varepsilon_n; n \in I\!N\}$ be a sequence of positive real numbers such that $\varepsilon_n \to 0^+$. We set

$$\Delta(\varepsilon_n) \; := \; \Delta(\psi + \varepsilon_n \parallel \cdot \parallel^2, C)$$
$$= \; \text{argmin}\{\psi(x) + \varepsilon_n \parallel x \parallel^2 | \, x \in C\}.$$

The following Lemma is well-known. We give its proof in order to be complete.

Lemma 3.1.1. Suppose that assumptions (h) are satisfied. If $\varepsilon > 0$ then $\Delta(\varepsilon) \neq \emptyset$.

Proof: The well-known direct method can be used for proving this classical result. Let us denote by $F_\varepsilon : X \to I\!R \cup \{+\infty\}$ the functional given by the formula

$$F_\varepsilon(x) \; := \; \psi(x) + \varepsilon \parallel x \parallel^2 .$$

We set

$$c \; := \; \inf\{F_\varepsilon(x) \mid x \in C\}.$$

Assumption (h_5) implies that $c < +\infty$ and assumption (h_4) ensures that $c > -\infty$. Then let $\{u_n; n \in I\!N\}$ be a minimizing sequence for the functional F_ε, i.e.

$$F_\varepsilon(u_n) \; \leq \; c + 1/n.$$

Using assumption (h_4), we obtain

$$c + 1/n \; \geq \; F_\varepsilon(u_n) \; \geq \; \varepsilon \parallel u_n \parallel^2 - \alpha_1 \parallel u_n \parallel^\beta - \alpha_2.$$

Therefore the sequence $\{u_n; n \in I\!N\}$ is bounded and we can take a subsequence (again denoted by u_n) such that $u_n \xrightarrow{\tau} u$ and

$$\psi(u_n) + \varepsilon \parallel u_n \parallel^2 \; \leq \; c + 1/n.$$

Using the τ-lower semicontinuity of ψ together with assumption (h_1), we get

$$F_\varepsilon(u) \; \leq \; c$$

and thus

$$F_\varepsilon(u) \leq F_\varepsilon(v), \forall v \in C.$$

∎

Our recession approach will be based on the properties of the following set of asymptotic directions:

$$R(\Delta(\varepsilon_n)) = \{w \in X \mid \exists u_n \in \Delta(\varepsilon_n), \| u_n \| \to +\infty$$
$$\text{and } w_n := u_n / \| u_n \| \xrightarrow{\tau} w\}.$$

Theorem 3.1.1. Suppose that assumptions (h) are satisfied. If

$$R(\Delta(\varepsilon_n)) = \emptyset.$$

then there exists $u \in C$ such that

$$\psi(u) = \min\{\psi(v) \mid v \in C\}$$

Proof: If $R(\Delta(\varepsilon_n)) = \emptyset$ then there exists $K > 0$ such that

$$\| z \| \leq K, \forall z \in \Delta(\varepsilon_n), \forall n \in \mathbb{N}.$$

Indeed, if we suppose the contrary, then we can find a sequence $\{z_n; n \in \mathbb{N}\}$ such that $\| z_n \| \to +\infty, z_n \in \Delta(\varepsilon_n)$ and thus for a subsequence, $z_n. \| z_n \|^{-1} \xrightarrow{\tau} w \in R(\Delta(\varepsilon_n))$, which is a contradiction.

Let $\{u_n; n \in \mathbb{N}\}$ be a sequence such that $u_n \in \Delta(\varepsilon_n)$. Then $\{u_n; n \in \mathbb{N}\}$ is bounded. For a subsequence, we can assume that $u_n \xrightarrow{\tau} u$ and for each $v \in C$, we have

$$\psi(u_n) + \varepsilon_n \| u_n \|^2 \leq \psi(v) + \varepsilon_n \| v \|^2 .$$

It is clear that

$$\psi(u_n) \leq \psi(v) + \varepsilon_n \| v \|^2 .$$

23

Therefore, by using the τ-lower semicontinuity of ψ, we obtain

$$\psi(u) \leq \psi(v), \forall v \in C.$$

∎

Now, we will point out a general process which can be used to prove that $R(\Delta(\varepsilon_n))$ is empty.

Firstly, on the set of asymptotic directions, we introduce a compactness condition by saying that $R(\Delta(\varepsilon n))$ is *asymptotically compact with respect to the topology* τ ($a(\tau)$-compact for short) if the following property holds true: If $\{w_n; n \in I\!N\}$ is a sequence such that

$$w_n := u_n/ \| u_n \| \xrightarrow{\tau} w,$$

with

$$u_n \in \Delta(\varepsilon_n)$$

and

$$\| u_n \| \to +\infty,$$

then there exists a subsequence (again denoted by w_n) such that $w_n \to w$ (convergence for the strong topology).

If $R(\Delta(\varepsilon_n))$ is asymptotically compact with respect to the weak topology, then we simply say (for short) that $R(\Delta(\varepsilon_n))$ is a-compact.

We are now in a position to prove several basic theoretical results which will be used later in this work.

For $\mu > 0$ given, we set

$$D_\mu := \{w \in X \mid u - \mu w \in C, \forall u \in C$$
$$\text{and } \psi(u - \mu w) \leq \psi(u), \forall u \in C\}$$

For instance, set $C = \mathbb{R}^N$ and let $A \in \mathbb{R}^{N \times N}$ be a positive semidefinite matrix. If $\psi(x) = \frac{1}{2}\langle Ax, x \rangle$ then it can be easily seen that

$$D_\mu = Ker(A + A^T), \ \forall \, \mu > 0$$

Set $C = \mathbb{R}$ and $\psi(x) = e^x$, then

$$D_\mu = \mathbb{R}_+, \forall \, \mu > 0.$$

Corollary 3.1.1. Suppose that assumptions (h) are satisfied. If

i) $X(= X')$ is a real Hilbert space;

ii) for all $w \in R(\Delta(\varepsilon_n))$ there exists $\mu(w) > 0$ such that

$$w \in D_{\mu(w)};$$

iii) $0 \notin R(\Delta(\varepsilon_n))$.

Then there exists $u \in C$ such that

$$\psi(u) = \min\{\psi(v) \mid v \in C\}.$$

Proof: We claim that $R(\Delta(\varepsilon_n))$ is empty. Indeed, if we suppose the contrary then we can find a sequence $u_n \in C$ such that $\| u_n \| \to +\infty, w_n := u_n / \| u_n \| \rightharpoonup w$, $u_n - \mu(w)w \in C$ and $\psi(u_n - \mu(w)w) \leq \psi(u_n)$. We have

$$\psi(u_n) + \varepsilon_n \| u_n \|^2 \leq \psi(u_n - \mu(w)w) + \varepsilon_n \| u_n - \mu(w)w \|^2$$
$$\leq \psi(u_n) + \varepsilon_n \| u_n - \mu(w)w \|^2$$

and thus

$$\| u_n \|^2 \leq \| u_n - \mu(w)w \|^2 .$$

We have also

$$\| u_n - \mu(w)w \|^2 = \| u_n \|^2 - 2\mu(w)\langle w, u_n \rangle + \| \mu(w)w \|^2$$

and thus

$$2\mu(w)\langle w, u_n \rangle \leq \mu(w)^2 \| w \|^2 . \tag{6}$$

Dividing (6) by $\| u_n \|$, we get

$$2\mu(w)\langle w, w_n \rangle \leq \mu(w)^2 \| w \|^2 \| u_n \|^{-1} .$$

Taking the limit as $n \to +\infty$, we obtain

$$2\mu(w) \| w \|^2 \leq 0$$

so that $w = 0$ and a contradiction to assumption iii).

∎

The following similar result holds true for a general real Banach space satisfying assumption (h_1).

Corollary 3.1.2. Suppose that assumptions (h) are satisfied. If

i) for all $w \in R(\Delta(\varepsilon_n))$, there exists $\mu(w) > 0$ such that

$$w \in D_{\mu(w)};$$

ii) $R(\Delta(\varepsilon_n))$ is $a(\tau)$-compact,

Then there exists $u \in C$ such that

$$\psi(u) = \min\{\psi(v) \mid v \in C\}.$$

Proof: We claim that $R(\Delta(\varepsilon_n))$ is empty. Indeed, if we suppose the contrary, then as in the previous Corollary, we can find a sequence $u_n \in C$ such that $\| u_n \| \to +\infty, w_n := u_n / \| u_n \| \xrightarrow{\tau} w, u_n - \mu(w)w \in C$ and

$$\| u_n \|^2 \leq \| u_n - \mu(w)w \|^2 . \tag{7}$$

26

However, we have also (for n large enough)

$$
\begin{aligned}
\| u_n - \mu(w)w \| & = \| u_n - \mu(w)u_n. \| u_n \|^{-1} + \mu(w)w_n - \mu(w)w \| \\
& = \| (1 - \mu(w) \| u_n \|^{-1})u_n + \mu(w)w_n - \mu(w)w \| \\
& \leq (1 - \mu(w) \| u_n \|^{-1}) \| u_n \| + \mu(w) \| w_n - w \| \\
& = \| u_n \| + \mu(w)(\| w_n - w \| - 1).
\end{aligned}
$$

Since by assumption ii), $w_n \to w$ (for a subsequence), we get

$$
\| u_n - \mu(w)w \| < \| u_n \|,
$$

for n large enough. This is a contradiction to (7).

$\blacksquare$

Remarks 3.1.1.

i) If $R(\Delta(\varepsilon_n))$ is $a(\tau)$-compact then $0 \notin R(\Delta(\varepsilon_n))$.

ii) Suppose that ψ is convex and continuous on $C = X$. Suppose also that $Z := R(\Delta(\varepsilon_n))$ is a closed subspace of X. If

$$
\partial\psi(u) \cap Z^+ \neq \emptyset, \forall u \in X
$$

(here Z^+ denotes the polar set of Z) then

$$
Z \subset D_1.
$$

Indeed,

$$
D_1 = \{w \in X \mid \psi(u) \leq \psi(u + w), \forall u \in X\}
$$

and by a result of Heron-Sermange ([81]; Lemma 1.2) $\partial\psi(u) \cap Z^+ \neq \emptyset$ if and only if $\psi(u) \leq \psi(u + z), \forall z \in Z$.

27

iii) Reciprocally, if

$$Z \subset D_1$$

then

$$\partial \psi(u) \cap Z^+ \neq \emptyset, \forall u \in X.$$

For a function $f : X \to \mathbb{R} \cup \{+\infty\}$, we denote by $K_o(f)$ the following level set

$$K_o(f) := \{x \in X : f(x) \leq 0\}.$$

The set $K_0(\tau - f_\infty)$ is called the τ-recession cone of f.

The following Proposition highlights the properties of the recession set $R(\Delta(\varepsilon_n))$ and it will be the key for our recession analysis.

Proposition 3.1.1. Suppose that assumptions (h) are satisfied. If $w \in R(\Delta(\varepsilon_n))$ then there exists a sequence $\{u_n; n \in \mathbb{N}\}$ such that

$\quad$ (1) $\quad u_n \in C$;

$\quad$ (2) $\quad \| u_n \| \to +\infty$;

$\quad$ (3) $\quad w_n := u_n \cdot \| u_n \|^{-1} \xrightarrow{\tau} w \in \tau - C_\infty \cap K_o(\tau - \psi_\infty)$;

$\quad$ (4) $\quad \limsup \psi(u_n)/ \| u_n \|^p \leq 0, \forall p > 0$;

$\quad$ (5) $\quad$ if $v_o \in C \cap \mathrm{dom}\{\psi\}$ then $\limsup \psi(u_n) \leq \psi(v_o)$.

If $R(\Delta(\varepsilon_n))$ is $a(\tau)$-compact and if $w \in R(\Delta(\varepsilon_n))$ then there exists a sequence $\{u_n; n \in \mathbb{N}\}$ satisfying (1), (2), (4), (5) and

$$(3') \quad w_n := u_n \cdot \| u_n \|^{-1} \to w \in C_\infty \cap K_o(\psi_\infty) \backslash \{0\}.$$

If for some $\mu > 0, R(\Delta(\varepsilon_n)) \subset D_\mu$ and if $w \in R(\Delta(\varepsilon_n))$ then there exists a sequence $\{u_n; n \in \mathbb{N}\}$ satisfying (1)-(5) and

$$(6) \quad \| u_n \| \leq \| u_n - \lambda w \|, \forall \lambda \geq \mu.$$

If $R(\Delta(\varepsilon_n))$ is $a(\tau)$-compact, $R(\Delta(\varepsilon_n)) \subset D_\mu$ for some $\mu > 0$ and if $w \in R(\Delta(\varepsilon_n))$ then there exists a sequence satisfying (1), (2), (4), (5), (6) and (3').

Proof: If $w \in R(\Delta(\varepsilon_n))$ then by Definition, there exists a sequence $\{u_n; n \in I\!\!N\}$ such that (1) $u_n \in C$, (2) $\| u_n \| \to +\infty$, (3) $w_n := u_n.\| u_n \|^{-1} \overset{\tau}{\to} w \in \tau - C_\infty$ and

$$\psi(u_n) + \varepsilon_n \| u_n \|^2 \le \psi(v) + \varepsilon_n \| v \|^2, \forall v \in C.$$

Let $v_o \in C \cap \mathrm{dom}\{\psi\}$ be given. We have

$$\psi(u_n) \le \psi(u_n) + \varepsilon_n \| u_n \|^2 \le \psi(v_o) + \varepsilon_n \| v_o \|^2 . \tag{8}$$

Therefore,

$$\limsup \psi(u_n) \le \psi(v_o)$$

and property (5) is satisfied.

Let $p > 0$ be given. Dividing (8) by $\| u_n \|^p$ and taking the limit as $n \to +\infty$, we get

$$\limsup \psi(u_n)/ \| u_n \|^p \le 0$$

and property (4) is satisfied. Inequality (8) implies also that

$$\liminf \psi(u_n)/ \| u_n \| \le 0$$

and thus $\tau - \psi_\infty(w) \le 0$.

If $R(\Delta(\varepsilon_n))$ is $a(\tau)$-compact then by considering a subsequence, we may assume that $w_n \to w$. Thus $w \in C_\infty$ and $\psi_\infty(w) \le 0$.

If for some $\mu > 0, R(\Delta(\varepsilon_n)) \subset D_\mu$ then as in Corollary 3.1.1 we prove that

$$\| u_n \| \le \| u_n - \mu w \| .$$

If $0 < t \le 1$ then

$$
\begin{aligned}
\| u_n \| \quad &\le \quad \| u_n - tu_n + tu_n - \mu w \| \\
&\le \quad (1 - t) \| u_n \| + t \| u_n - \mu t^{-1} w \|
\end{aligned}
$$

29

so that

$$\| u_n \| \leq \| u_n - \mu t^{-1} w \| .$$

Therefore

$$\| u_n \| \leq \| u_n - \lambda w \|, \forall \lambda \geq \mu.$$

∎

In practice, we do not check the compactness condition on the set of sequences defined by $R(\Delta(\varepsilon_n))$ but on a larger set of sequences chosen with the help of Proposition 3.1.1. For instance, we may examine the set of sequences $\{u_n, n \in I\!N\}$ satisfying (1)-(4) of Proposition 3.1.1.

Remarks 3.1.2. By using Proposition 3.1.1 together with Theorem 3.1.1, it is now easy to obtain some basic conditions guaranteeing the existence of a minimum for ψ on C. For instance:

i) If ψ is coercive, i.e.

$$\psi(u) \rightarrow +\infty \text{ as } \| u \| \rightarrow +\infty, u \in C$$

then $R(\Delta(\varepsilon_n)) = \emptyset$. Indeed, if we suppose the contrary, then we get a contradiction to Proposition 3.1.1 (5).

ii) If $R(\Delta(\varepsilon_n))$ is $a(\tau)$-compact and $C_\infty \cap K_o(\psi_\infty) = \{0\}$ then $R(\Delta(\varepsilon_n)) = \emptyset$. Indeed, if we assume the contrary then we obtain a contradiction to Proposition 3.1.1(3').

iii) If C is bounded then $R(\Delta(\varepsilon_n)) = \emptyset$. If we suppose the contrary then we contradict Proposition 3.1.1(2).

iv) If $R(\Delta(\varepsilon_n))$ is $a(\tau)$-compact and $\psi_\infty(x) > 0, \forall x \in X\backslash\{0\}$, then $R(\Delta(\varepsilon_n)) = \emptyset$. In this case $K_0(\psi_\infty) = \{0\}$ and we may conclude as in Remark ii).

30

v) If $\dim(X) < +\infty$ then the a-compacity of $R(\Delta(\varepsilon_n))$ is automatically satisfied.

vi) If assumptions (h_1) and (h_2) are satisfied then it is easy to see that if $\tau - \psi_\infty(x) \geq 0, \forall x \in X$, then assumption (h_4) is satisfied too. Otherwise, we can find a sequence $x_n \in X$ such that

$$\psi(x_n) < - \| x_n \| - n.$$

The sequence $\{x_n, n \in \mathbb{N}\}$ is unbounded. Indeed, if we suppose that $\{x_n, n \in \mathbb{N}\}$ is bounded, then we may take a subsequence satisfying $x_n \xrightarrow{\tau} x \in X$. Using (h_3) we get $\psi(x) \leq -\infty$ which is a contradiction. Setting $t_n := \| x_n \|$ and $y_n := \frac{x_n}{t_n}$, we obtain for a subsequence, $y_n \xrightarrow{\tau} y, t_n \to +\infty$ and

$$\psi(t_n y_n)/t_n < -1 - n t_n^{-1}.$$

Thus $\tau - \psi_\infty(y) < 0$ which is a contradiction.

vii) If $\tau - \psi_\infty(x) \geq 0, \forall x \in X$, then the perturbed function $x \to \psi(x) + \varepsilon \| x \|^2$ ($\varepsilon > 0$) is coercive, i.e. there exists $\alpha > 0$ and $R > 0$ such that

$$\psi(x) + \varepsilon \| x \|^2 \geq \alpha \| x \|^2, \forall x \in X, \| x \| \geq R.$$

Indeed, if we suppose the contrary, then there exists a sequence $\{x_n; n \in \mathbb{N}\}$ such that

$$\| x_n \| \to +\infty$$

and

$$\psi(x_n) < (n^{-1} - \varepsilon) \| x_n \|^2 .$$

Then, for a subsequence, we can assume that

$$w_n := x_n / \| x_n \| \xrightarrow{\tau} w$$

31

and

$$t_n := \| x_n \| \to +\infty.$$

Therefore (for n great enough)

$$\psi(t_n w_n)/t_n < (n^{-1} - \varepsilon)t_n \leq -\frac{\varepsilon}{2}t_n,$$

so that

$$0 \leq \tau - \psi_\infty(x) \leq -\infty,$$

which is a contradiction. Note also that assumption (h_4) in our previous results could be replaced by the coercivity of the perturbed function $x \to \Psi(x) + \varepsilon \| x \|^2$, for each $\varepsilon > 0$.

viii) If ψ is a proper, convex and lower-semicontinuous function then assumption (h_4) is satisfied [122].

In many applications, it will be often easy to use Proposition 3.1.1 so as to check the $a(\tau)$-compacity of $R(\Delta(\varepsilon_n))$ and to exhibit a subset N containing $R(\Delta(\varepsilon_n))$ and such that the inequality $\psi_\infty(x) > 0$ is satisfied on $N\backslash\{0\}$. The following result is in this sense.

Corollary 3.1.3. Suppose that assumptions (h) are satisfied. If there exists a subset $N \subset X$ such that

i) $R(\Delta(\varepsilon_n)) \subset N\backslash\{0\}$;

ii) $\psi_\infty(x) > 0, \forall x \in N\backslash\{0\}$;

and

iii) $R(\Delta(\varepsilon_n))$ is $a(\tau)$-compact.

Then there exists $u \in C$ such that

$$\psi(u) \ = \ \min\{\psi(v) \mid v \in C\}.$$

For instance, as a consequence of Corollary 3.1.3, we obtain the following basic result.

Corollary 3.1.4. Suppose that

i) X is a real Hilbert space;

ii) C is a weakly closed subset of X;

iii) $A : X \to X'$ is a bounded linear symmetric and semicoercive operator;

iv) $\dim\{Ker(A)\} < +\infty$;

v) $F : X \to \mathbb{R} \cup \{+\infty\}$ is a weakly lower semicontinuous functional such that $\text{dom}\{F\} \cap C \neq \emptyset$;

vi) there exist $\alpha_1 \geq 0, \alpha_2 \in \mathbb{R}$ and $0 < \beta < 2$ such that

$$F(x) \ \geq \ -\alpha_1 \parallel x \parallel^\beta -\alpha_2, \forall x \in X;$$

vii) $f \in X'$.

Let us denote by $\psi : X \to \mathbb{R} \cup \{+\infty\}$ the functional

$$\psi(u) \ := \ \tfrac{1}{2}\langle Au, u\rangle + F(u) - \langle f, u\rangle.$$

A) If

$$F_\infty(e) > \langle f, e\rangle, \forall e \in Ker(A) \cap C_\infty, e \neq 0, \tag{9}$$

then there exists $u \in C$ such that

$$\psi(u) \ \leq \ \psi(v), \forall v \in C.$$

B) Suppose in addition that the function $x \rightarrow F(x)$ and the set C are convex. Then if there exists a minimizer for ψ on C then

$$F_\infty(e) \geq \langle f, e \rangle, \forall e \in Ker(A) \cap C_\infty. \tag{10}$$

Moreover, if u_1 and u_2 are two minimizers for ψ on C then

$$u_2 - u_1 \in Ker(A)$$

and

$$\langle f, u_2 - u_1 \rangle = F(u_2) - F(u_1).$$

Proof: Let $w \in R(\Delta(\epsilon_n))$ be given. By Proposition 3.1.1, there exists a sequence $u_n \in C$ such that $t_n := \| u_n \| \rightarrow +\infty, w_n := u_n/t_n \rightharpoonup w \in w - C_\infty$ and

$$\limsup\{\langle Aw_n, w_n \rangle + F(u_n)/ \| u_n \|^2 - \langle f, u_n \rangle/ \| u_n \|^2\} \leq 0.$$

Therefore

$$\liminf\langle Aw_n, w_n \rangle + \liminf(-\alpha_1 \| u_n \|^{\beta-2} -\alpha_2 \| u_n \|^{-2}) \leq 0$$

and thus

$$\liminf\langle Aw_n, w_n \rangle \leq 0.$$

Thus

$$w \in Ker(A)$$

and

$$(I - Q)w_n \rightarrow 0.$$

Here $Q : X \rightarrow Ker(A)$ denotes the orthogonal projector of X onto $Ker(A)$. Moreover, $Qw_n \rightarrow Qw$ since Q is bounded linear (and thus weakly continuous) and $\dim\{Ker(A)\} < +\infty$. We have

$$w_n = (I - Q)w_n + Qw_n \rightarrow 0 + Qw = w \in C_\infty.$$

34

Thus assumption i) and iii) of Corollary 3.1.3 are satisfied (here $N = Ker(A) \cap C_\infty$).

Using Proposition 2.1.1 ((2) and (4)), we have

$$\psi_\infty(x) \geq \psi_{Ker(A)}(x) + F_\infty(x) - \langle f, x \rangle.$$

Therefore assumption i) of Corollary 3.1.3 is satisfied if

$$F_\infty(e) > \langle f, e \rangle, \forall e \in Ker(A) \cap C_\infty \backslash \{0\}.$$

If F is convex and u is a minimizer for ψ on C then $u \in C$ and satisfies the following variational inequality

$$\langle Au, v - u \rangle + F(v) - F(u) \geq \langle f, v - u \rangle, \forall v \in C. \tag{11}$$

Thus

$$F(u + e) - F(u) \geq \langle f, e \rangle, \forall e \in Ker(A) \cap C_\infty.$$

Using Proposition 2.1.2 (2), we obtain

$$F_\infty(e) \geq \langle f, e \rangle, \forall e \in Ker(A) \cap C_\infty.$$

Using (11), it is easy to prove that

$$\langle Au_2 - Au_1, u_2 - u_1 \rangle \leq 0,$$

so that $u_2 - u_1 \in Ker(A)$. Moreover, we have

$$\tfrac{1}{2}a(u_1, u_1) + F(u_1) - \langle f, u_1 \rangle = \tfrac{1}{2}a(u_2, u_2) + F(u_2) - \langle f, u_2 \rangle$$

and thus

$$\langle f, u_2 - u_1 \rangle = F(u_2) - F(u_1).$$

■

35

Remarks 3.1.3.

i) Suppose that the function $x \to F(x)$ and the set C are convex. Moreover if we assume that

$$(a) \quad F(u + z) = F(u), \forall z \in Ker(A), u \in C;$$

$$(b) \quad F_\infty(e) > \langle f, e \rangle, \forall e \in Ker(A) \cap C_\infty \backslash \{0\}$$

and

$$(c) \quad \langle f, e \rangle \neq 0, \forall e \in Ker(A) \backslash \{0\}.$$

Then the functional $x \to \psi(x)$ defined in Corollary 3.1.4 admits a unique minimizer on C. Indeed, the existence is the result of condition (b) and Corollary 3.1.4. If we suppose the existence of two different minimizers u_1 and u_2 then using Corollary 3.1.4 together with condition (a), we get $u_2 - u_1 \in Ker(A) \backslash \{0\}$ and $\langle f, u_2 - u_1 \rangle = 0$ which is a contradiction to condition (c).

ii) For instance, let Ω be a bounded, open, connected subset of $I\!\!R^N$ with a Lipschitz boundary. If we suppose that $f \in L^2(\Omega)$ and

$$\int_\Omega f(x) dx > 0$$

then the problem of minimizing the functional

$$\psi(u) := \frac{1}{2} \int_\Omega |\nabla u(x)|^2 \, dx - \int_\Omega f(x) u(x) dx$$

on the set

$$C := \{u \in H^1(\Omega) \mid u(x) \geq 0 \text{ a.e. on } \Omega\}$$

admits a unique solution.

iii) Let us now briefly discuss the physical meaning of a condition like (10) in Corollary 3.1.4 or a condition like i) in Corollary 3.1.2. Consider a body which in its undeformed state occupies a bounded, open and connected domain of $I\!R^3$. Suppose that the energy of the deformation of the body is given by a semicoercive and symmetric bilinear form $a(u,v) = \langle Au, v \rangle$ which is such that the space of rigid body motions is

$$Ker A = \{v(x) = \alpha + \beta \wedge x, \alpha, \beta \in I\!R^3, x \in \Omega\}.$$

We suppose also that the body is subjected to body forces f and surface forces t. Finally, we assume that the displacement field u of the body is subjected to geometric constraints characterized by the relation $u \in C$, where C is a nonempty, closed and convex set. We set

$$\Psi(u) := \tfrac{1}{2}a(u,v) - \int_\Omega f.u \; dx - \int_{\partial\Omega} t.u \; ds.$$

We will show that suitable load conditions can be posed so as to check the abstract conditions mentioned here above (in this Remark, we suppose that the data are smooth enough to justify the following computations). Suppose that there exists $x_0 \in \Omega$ such that

$$\int_\Omega (x - x_0) \wedge f \; dx + \int_{\partial\Omega} (x - x_0) \wedge t \; ds = 0.$$

Then if $v \in Ker A$, we can write $v(x) = a + b \wedge (x - x_0)$ for suitable $a, b \in I\!R^3$ and:

$$
\begin{aligned}
\int_\Omega f.v \; dx + \int_{\partial\Omega} t.v \; ds &= \int_\Omega f.a \; dx + \int_{\partial\Omega} t.a \; ds + \int_\Omega (b \wedge (x - x_0)).f \; dx \\
&\quad + \int_{\partial\Omega} (b \wedge (x - x_0)).t \; ds \\
\\
&= \int_\Omega f.a \; dx + \int_{\partial\Omega} t.a \; ds.
\end{aligned}
$$

Let us now consider two examples of set of geometric constraints. a) We set

$$C = \{u : u.e_1 \geq 0, u.e_2 \geq 0 \text{ and } u.e_3 \geq 0\},$$

where $e_i (i = 1, 2, 3)$ are the canonical vectors of $\mathbb{R}^3$. Thus if $x \to v(x) = a + b \wedge (x - x_0) \in Ker A \cap C_\infty$, we obtain the inequalities

$$a.e_i + b \wedge (x - x_0).e_i \geq 0, \forall x \in \Omega, (i = 1, 2, 3).$$

Since $x_0 \in \Omega$ and if $v \neq 0$, we obtain $a.e_i \geq 0 (i = 1, 2, 3)$ and $a \neq 0$. We have

$$\int_\Omega f.a \; dx + \int_{\partial \Omega} t.a \; ds = \sum_{i=1}^{3} a.e_i [\int_\Omega f.e_i \; dx + \int_{\partial \Omega} t.e_i \; ds],$$

and thus if we assume in addition that

$$e_i.(\int_\Omega f \; dx + \int_{\partial \Omega} t \; dx) < 0 \quad (i = 1, 2, 3)$$

then clearly

$$\int_\Omega f.v \; dx + \int_{\partial \Omega} t.v \; ds < 0, \forall v \in Ker A \cap C_\infty \backslash \{0\}.$$

b) We set

$$C = \{u : u.e_3 \geq 0\}.$$

Here, if $x \to v(x) = a + b \wedge (x - x_0) \in Ker A \cap C_\infty$, we obtain

$$a.e_3 \geq 0.$$

If we assume now that

$$e_1.(\int_\Omega f \; dx + \int_{\partial \Omega} t \; ds) = 0,$$
$$e_2.(\int_\Omega f \; dx + \int_{\partial \Omega} t \; ds) = 0$$

and

$$e_3.(\int_\Omega f \; dx + \int_{\partial \Omega} t \; ds) < 0,$$

then clearly

$$\int_\Omega f.v \; dx + \int_{\partial \Omega} t.v \; ds \leq 0, \forall v \in Ker A \cap C_\infty.$$

If we consider now $x \to v(x) = a + b \wedge (x - x_0)$ in the set $Ker A \cap C_\infty \cap K_0(\Psi_\infty) \backslash \{0\}$ then we obtain also the relation

$$\int_\Omega f.v \; dx + \int_{\partial \Omega} t.v \; ds \geq 0,$$

38

and thus

$$\sum_{i=1}^{3} a.e_i(\int_{\Omega} f.e_i \, dx + \int_{\partial\Omega} t.e_i \, ds) \geq 0.$$

Therefore $a.e_3 = 0$ and it follows that $(b\wedge(x-x_0)).e_3 \geq 0, \forall x \in \Omega$, or equivalently

$$b_1(x_2 - x_{0,2}) - a_2(x_1 - x_{0,1}) \geq 0, \forall x \in \Omega.$$

Since $x_0 \in \Omega$ and Ω is open, then this last relation implies that $b_1 = a_2 = 0$ and thus $(b \wedge (x - x_0)).e_3 = 0, \forall x \in \Omega$. Thus, for each $u \in C$ and for each $\mu > 0$, we obtain

$$(u - \mu v).e_3 = u \in C$$

and

$$\begin{aligned} \Psi(u - \mu v) &= \Psi(u) + \mu \sum_{i=1}^{3} a.e_i(\int_{\Omega} f.e_i \, dx + \int_{\partial\Omega} t.e_i \, ds) \\ &= \Psi(u). \end{aligned}$$

We end this Section with basic results connected to the energy functionals of class (Q), (Q^+) and (A) defined in Section 2.3.

Corollary 3.1.5. Suppose that assumptions (h) are satisfied. If

i) X is a real Hilbert space;

ii) $R(\Delta(\varepsilon_n)) \subset D_1$;

iii) ψ satisfies property (Q) on C.

Then there exists $u \in C$ such that

$$\psi(u) = \min\{\psi(v) \mid v \in C\}.$$

Proof: We claim that $0 \notin R(\Delta(\varepsilon_n))$. Indeed, if we suppose the contrary then we get a contradiction to Proposition 3.1.1 (5). We conclude by application of Corollary 3.1.1. ∎

Corollary 3.1.6. Suppose that assumptions (h) are satisfied. If

i) X is a real reflexive Banach space;

ii) $R(\Delta(\varepsilon_n)) \subset D_1$;

iii) ψ satisfies property (Q^+) on C.

Then there exists $u \in C$ such that

$$\psi(u) = \min\{\psi(v) \mid v \in C\}.$$

Proof: We claim that $R(\Delta(\varepsilon_n))$ is empty. Indeed, if we suppose the contrary then we get a contradiction to Proposition 3.1.1 (5). We conclude by application of Theorem 3.1.1.

∎

Reinforcing the compactness condition and using the concept of property (A) due to A. Auslender [18], we can refine condition i) in Corollary 3.1.2 to obtain a condition that is necessary and sufficient. We say that the compactness condition $(\mathbf{C})$ is satisfied if for each sequence $\{u_n, n \in I\!N\}$ satisfying

$$u_n \in C,$$
$$\|u_n\| \to +\infty,$$
$$w_n := u_n / \|u_n\| \overset{\tau}{\to} w$$

and

$$\limsup \psi(u_n) / \|u_n\|^P \leq 0, \forall\, p > 0,$$

there exists a subsequence such that $w_n \to w$.

Corollary 3.1.7. Suppose that assumptions (h) are satisfied. We assume that the compactness condition $(\mathbf{C})$ holds true. Then there exists $u \in C$ such that

$$\psi(u) = \min\{\psi(v) \mid v \in C\}$$

40

if and only if ψ has property (A) on C.

Proof: Suppose that ψ has property (A) on C. We claim that $R(\Delta(\varepsilon_n))$ is empty. Indeed, if we suppose the contrary, then by using Proposition 3.1.1 and Definition 2.3.4, we can find sequences $\{u_n, n \in I\!N\}, \{z_n, n \in I\!N\}$ and $\{\mu_n; n \in I\!N\}$ satisfying $u_n \in \Delta(\varepsilon_n)$, $\|u_n\| \to +\infty$, $0 < \mu_n \leq \|u_n\|$, $w_n := \frac{u_n}{\|u_n\|} \to w$, $z_n \to z$, $\|z - w\| < 1$, $u_n - \mu_n z_n \in C$ and $\psi(u_n - \mu_n z_n) \leq \psi(u_n)$. We have,

$$
\begin{aligned}
\psi(u_n) + \varepsilon_n \|u_n\|^2 &\leq \psi(u_n - \mu_n z_n) + \varepsilon_n \|u_n - \mu_n z_n\|^2 \\
&\leq \psi(u_n) + \varepsilon_n \|u_n - \mu_n z_n\|^2
\end{aligned}
$$

Thus

$$\|u_n\|^2 \leq \|u_n - \mu_n z_n\|^2 . \tag{12}$$

On the other hand, we have

$$
\begin{aligned}
\|u_n - \mu_n z_n\| &\leq (1 - \mu_n/\|u_n\|)\|u_n\| + \mu_n \|w_n - z_n\| \\
&= \|u_n\| + \mu_n(\|w_n - z_n\| - 1).
\end{aligned}
$$

Using (12) we obtain $\|w_n - z_n\| \geq 1$ and by taking the limit as $n \to +\infty$, it results that $\|w - z\| \geq 1$, which is a contradiction. We conclude by application of Theorem 3.1.1. Suppose now the existence of $u \in C$ satisfying

$$\psi(u) = \min\{\psi(v) \mid v \in C\},$$

and let $\{w_n; n \in I\!N\}$ be a sequence such that $w_n := \frac{u_n}{\|u_n\|} \xrightarrow{\tau} w$, with $u_n \in C$, $\|u_n\| \to +\infty$ and $\limsup \psi(u_n)/\|u_n\|^p \leq 0, \forall p > 0$. Using condition $(\mathbf{C})$, we take a subsequence such that $w_n \to w$. We set $z_n := \frac{u_n - u}{\|u_n\|}$ and $\mu_n = \|u_n\|$. We see that

$$u_n - \mu_n z_n = u \in C,$$
$$\psi(u_n - \mu_n z_n) \leq \psi(u_n),$$

and

$$z_n \to w.$$

41

Thus ψ has property (A) on C.　　　　　　　　　　　　　　　　　∎

Remarks 3.1.4. i) If condition (C) in Corollary 3.1.7 is replaced by the $a(\tau)$–compactness of $R(\Delta(\varepsilon_n))$ then the hypothesis requiring that ψ has property (A) on C remains a sufficient condition for the existence of a minimizer of ψ on C. ii) In Remark 2.3.2, we have seen that the function

$$\psi(x) = \begin{cases} e^{-x^2} & \text{if } x \in (-\infty, -1) \cup (+1, +\infty) \\ \mid (1 + e^{-1})x \mid -1 & \text{if } x \in [-1, +1] \end{cases}$$

has property (A) on $\mathbb{R}$. This function satisfies assumptions (h) too. Note that it is not possible to find $\mu > 0$ such that $1 \in D_\mu$. Indeed, if we suppose the contrary, we would obtain $e^{2\mu u} \leq e^{\mu^2}$, for all $u \geq 1 + \mu$. This is a contradiction for u chosen sufficiently large. That means that Corollary 3.1.2 cannot be applied. iii) We say that ψ has property (A') on C if there exists $K > 0$ such that

$$\forall y \in X, \exists z \in X, \exists \mu \in (0, \| y \|] \text{ such that } y - \mu z \in C,$$
$$f(y - \mu z) \leq f(y) \text{ and } \| \tfrac{y}{\|y\|} - z \| \leq \tfrac{K}{\|y\|}.$$

If ψ satisfies condition (C) and has property (A') on C, then ψ has property (A) on C. Indeed, let $\{u_n, n \in \mathbb{N}\}$ be a sequence such that $u_n \in C, \| u_n \| \to +\infty$, $w_n := u_n / \| u_n \| \xrightarrow{\tau} w$ and $\limsup \psi(u_n) / \| u_n \|^p \leq 0, \forall p > 0$. Using condition (C) we can consider a subsequence such that $w_n \to w$. Using now property (A') we find sequences $\{\mu_n; n \in \mathbb{N}\}$ and $\{z_n; n \in \mathbb{N}\}$ such that $0 < \mu_n \leq \| u_n \|, u_n - \mu_n z_n \in C$, $\psi(u_n - \mu_n z_n) \leq \psi(u_n)$ and $\| w_n - z_n \| \leq \tfrac{K}{\|u_n\|}$. Thus $z_n \to w$ and ψ satisfies property (A). Note also that if ψ has a minimum on C then ψ has property (A'). Indeed, let u such a minimizer. For $y \in X$, we set $K = \| u \|, z = \tfrac{y-u}{\|y\|}$ and $\mu = \| y \|$. We have $y - \mu z = u \in C, \psi(y - \mu z) \leq \psi(y), \| \tfrac{y}{\|y\|} - z \| = \tfrac{K}{\|y\|}$. Therefore ψ has property (A'). iv) If X is a real Hilbert space, then we can also modify the concept of (A)–property so as to refine condition ii) in Corollary 3.1.1. We say that ψ has property (A^+) on C

if the following holds true: If $\{u_n; n \in I\!N\}$ is a sequence such that

$$u_n \in C,$$

$$\| u_n \| \to +\infty,$$

$$w_n := u_n/ \| u_n \| \rightharpoonup w,$$

$$\limsup \psi(u_n)/ \| u_n \|^p \le 0, \forall p > 0,$$

then there exists a subsequence (again denoted by $\{u_n, n \in I\!N\}$), a real number $K > 0$ and sequences $\{z_n, n \in I\!N\} \subset X$ and $\{\mu_n, n \in I\!N\}$ such that

$$0 < \mu_n \le K,$$

$$z_n \to z,$$

$$\langle w, z \rangle > 0,$$

$$u_n - \mu_n z_n \in C$$

and

$$\psi(u_n - \mu_n z_n) \le \psi(u_n).$$

Corollary 3.1.8. Suppose that assumptions (h) are satisfied. Moreover, we assume that

1) X is a real Hilbert space;

ii) $0 \notin R(\Delta(\epsilon_n))$;

iii) ψ has property (A^+) on C.

Then there exists $u \in C$ such that

$$\psi(u) = \min\{\psi(v) \mid v \in C\}.$$

Proof: Suppose that the (A^+)–condition is satisfied. Then we claim that assumptions ii) and iii) imply that $R(\Delta(\epsilon_n))$ is empty. Indeed, if we suppose the contrary then we can find sequences $\{u_n, n \in I\!N\}$, $\{z_n; n \in I\!N\}$ and $\{\mu_n; n \in I\!N\}$ satisfying $u_n \in C$,

43

$\| u_n \| \to +\infty, \, 0 < \mu_n \le K, w_n = \frac{u_n}{\|u_n\|} \rightharpoonup w, \, z_n \to z, \, \langle w, z \rangle > 0, u_n - \mu_n z_n \in C$

and $\psi(u_n - \mu_n z_n) \le \psi(u_n)$. As in Corollary 3.1.1, we see that

$$\| u_n \|^2 \le \| u_n - \mu_n z_n \|^2 .$$

We have also

$$\| u_n - \mu_n z_n \|^2 = \| u_n \|^2 - 2\mu_n \langle z_n, u_n \rangle + \mu_n^2 \| z_n \|^2,$$

and thus

$$2\langle z_n, u_n \rangle \le K \| z_n \|^2 .$$

Dividing this last relation by $\| u_n \|$ and taking the limit as $n \to +\infty$, we obtain

$$\langle z, w \rangle \le 0,$$

which is a contradiction. ∎

3.2 Recession analysis involving a general viscosity approach

The concept of viscosity solutions plays a central role in the study of many problems in mathematics: Calculus of variations, Hamilton-Jacobi equations, singular perturbations and optimal control theory. See for instance [11], [49], [133] and the references cited therein. The viscosity methods for minimization problems have been recently reviewed and formulated in a general form by H. Attouch in [12].

We suppose that the following assumptions (j) hold true.

(j_1) X is a Banach space such that either X is reflexive or $X = V'$ with V separable and τ is the weak* topology of X.

(j_2) $\psi : X \to \mathbb{R} \cup \{+\infty\}; x \to \psi(x)$ is τ-lower semicontinuous;

(j_3) $C \subset X$ is τ-closed and $\text{dom}\{\psi\} \cap C \neq \emptyset$.

(j_4) $g : X \rightarrow I\!R$ is a nonnegative real-valued and τ-lower semicontinuous functional.

The function $g : X \rightarrow I\!R^+$ is called the viscosity function [12] and is assumed to enjoy the property that

(j_5) for each $\varepsilon > 0$,

$$\operatorname{argmin}\{\psi(x) + \varepsilon g(x) \mid x \in C\} \neq \emptyset.$$

Let $\{\varepsilon_n; n \in I\!N\}$ be a sequence of positive real numbers such that $\varepsilon_n \rightarrow 0^+$.

We set

$$\begin{aligned}
\Delta(\varepsilon_n, g) \quad &:= \quad \Delta(\psi + \varepsilon_n g, C) \\
&= \quad \operatorname{argmin}\{\psi(x) + \varepsilon_n g(x) \mid x \in C\}.
\end{aligned}$$

Theorem 3.2.1. Suppose that assumptions (j) are satisfied. If

$$R(\Delta(\varepsilon_n, g)) \;=\; \emptyset.$$

Then there exists $u \in C$ such that

$$\psi(u) \;=\; \min\{\psi(v) \mid v \in C\}.$$

Moreover

$$g(u) \;\leq\; g(v), \forall v \in \operatorname{argmin}\{\psi(x) \mid x \in C\}. \tag{13}$$

Proof: As in Theorem 3.1.1, we prove that if $R(\Delta(\varepsilon_n, g)) \;=\; \emptyset$ then we can find a bounded sequence $u_n \in \Delta(\varepsilon_n, g)$. Then as in Theorem 3.1.1, we obtain the existence of $u \in C$ such that $u_n \xrightarrow{\tau} u$ and

$$\psi(u) \;\leq\; \psi(v), \forall v \in C.$$

Moreover, we have

$$\varepsilon_n^{-1}(\psi(u_n) - \inf\{\psi(x) \mid x \in C\}) + g(u_n) \;\leq\; \varepsilon_n^{-1}(\psi(v) - \inf\{\psi(x) \mid x \in C\}) + g(v).$$

45

Thus if $v \in \text{argmin}\{\psi(x) \mid x \in C\}$ then we get

$$g(u_n) \leq g(v)$$

and then

$$g(u) \leq g(v).$$

∎

Now, we are able to characterize the solution obtained by the recession approach as the one satisfying a given selection principle.

Corollary 3.2.1. Suppose that assumptions (j) are satisfied. If for some $m > 0$

i) g is τ-continuous and satisfies

$$g^m(\alpha x + \beta y) \leq \alpha g^m(x) + \beta g^m(y), \ \forall \, x, \, y \in X, \, \forall \, \alpha, \, \beta > 0;$$

ii) there exists $c > g^m(0)$ and $R > 0$ such that

$$x \in C, \| x \| \geq R \Rightarrow g^m(x) \geq c \| x \|;$$

iii) for all $w \in R(\Delta(\varepsilon_n, g))$, there exists $\mu(w) > 0$ such that

$$w \in D_{\mu(w)};$$

iv) $R(\Delta(\varepsilon_n, g))$ is $a(\tau)$-compact.

Then there exists $u \in C$ such that

$$\psi(u) = \min\{\psi(v) \mid v \in C\}$$

and

$$g(u) \leq g(v), \forall v \in \text{argmin}\{\psi(x) \mid x \in C\}.$$

Proof: We claim that $R(\Delta(\varepsilon_n))$ is empty. Indeed, if we suppose the contrary then we can find a sequence $u_n \in C$ such that $\| u_n \| \to +\infty$, $w_n := u_n / \| u_n \| \xrightarrow{\tau} w$, $u_n - \mu(w)w \in C$ and $\psi(u_n - \mu(w)w) \leq \psi(u_n)$.

We have

$$\psi(u_n) + \varepsilon_n g(u_n) \leq \psi(u_n - \mu(w)w) + \varepsilon_n g(u_n - \mu(w)w)$$
$$\leq \psi(u_n) + \varepsilon_n g(u_n - \mu(w)w)$$

and thus since g is nonnegative

$$g^m(u_n) \leq g^m(u_n - \mu(w)w). \tag{14}$$

On the other hand, we have also

$$g^m(u_n - \mu(w)w) = g^m(u_n - \mu(w)u_n \| u_n \|^{-1} + \mu(w)w_n - \mu(w)w)$$
$$= g^m((1 - \mu(w) \| u_n \|^{-1})u_n + \mu(w)(w_n - w)).$$

If n is great enough then we can use assumption i) in order to get

$$g^m(u_n - \mu(w)w) \leq (1 - \mu(w). \| u_n \|^{-1})g^m(u_n) + \mu(w)g^m(w_n - w)$$
$$= g^m(u_n) + \mu(w)(g^m(w_n - w) - g^m(u_n) \| u_n \|^{-1}).$$

By $a(\tau)$-compacity, we have $w_n \to w$ (for a subsequence), and assumption i) together with assumption ii) imply that $g^m(w_n - w) - g^m(u_n) \| u_n \|^{-1} < 0$, for n great enough. Thus

$$g^m(u_n - \mu(w)w) < g^m(u_n),$$

for n great enough. That is a contradiction to (14).

■

Remarks 3.2.1.

i) Inequality (13) is called the viscosity selection principle [12].

47

ii) If we set $g(x) := \| x \|^2$ then we recover the Tychonov regularization. Here, the selection principle is a consequence of the recession approach involving a regularization of the original function ψ.

It is now easy to show that Proposition 3.1.1 remains true with assumptions (j).

Proposition 3.2.1. Suppose that assumptions (j) are satisfied. If $w \in R(\Delta(\varepsilon_n, g))$ then there exists a sequence $\{u_n; n \in I\!N\}$ such that

(1) $u_n \in C$;

(2) $\| u_n \| \to +\infty$;

(3) $w_n := u_n \cdot \| u_n \|^{-1} \xrightarrow{\tau} w \in \tau - C_\infty \cap K_o(\tau - \psi_\infty)$;

(4) $\limsup \psi(u_n)/ \| u_n \|^p \le 0, \forall p > 0$;

(5) if $v_o \in C \cap \mathrm{dom}\{\psi\}$ then $\limsup \psi(u_n) \le \psi(v_o)$.

If $R(\Delta(\varepsilon_n, g))$ is $a(\tau)$-compact and if $w \in R(\Delta(\varepsilon_n, g))$ then there exists a sequence $\{u_n; n \in I\!N\}$ satisfying (1), (2), (4), (5) and

(3') $w_n := u_n \cdot \| u_n \|^{-1} \to w \in C_\infty \cap K_o(\psi_\infty) \backslash \{0\}$.

Suppose in addition that g is convex. Then, if for some $\mu > 0, R(\Delta(\varepsilon_n, g)) \subset D_\mu$ and if $w \in R(\Delta(\varepsilon_n), g)$ then there exists a sequence $\{u_n; n \in I\!N\}$ satisfying (1)-(5) and

(6) $g(u_n) \le g(u_n - \lambda w), \forall \lambda \ge \mu$

If $R(\Delta(\varepsilon_n, g))$ is $a(\tau)$-compact, $R(\Delta(\varepsilon_n, g)) \subset D_\mu$ for some $\mu > 0$ and if $w \in R(\Delta(\varepsilon_n), g)$ then there exists a sequence satisfying (1), (2), (4), (5), (6) and (3').

Proof: (1-5) and (3') are proved by following similar arguments than in Proposition 3.1.1. It remains to prove (6). If for some $\mu > 0, R(\Delta(\varepsilon_n, g)) \subset D_\mu$ then

$$g(u_n) \le g(u_n - \mu w).$$

48

If $0 < t \leq 1$ then

$$
\begin{aligned}
g(u_n) &\leq g(u_n - tu_n + tu_n - \mu w) \\
&\leq (1-t)g(u_n) + tg(u_n - \mu t^{-1}w)
\end{aligned}
$$

so that

$$
g(u_n) \leq g(u_n - \mu t^{-1}w).
$$

■

3.3 On the recession analysis for nonvariational problems on reflexive Banach spaces

Let C be a nonempty subset of a real Banach space X. If u is a minimizer for $\Phi \in C^1(X; \mathbb{R})$ on C then $u \in C$ and

$$
\langle \Phi'(u), v \rangle \geq 0, \forall v \in T_C(u) \tag{15}
$$

where $T_C(u)$ denotes Clarke's tangent cone of C at $u \in C$. Our aim in this chapter is now to consider the nonvariational problem: Find $u \in C$ such that

$$
\langle Au - f, v \rangle \geq 0, \forall v \in T_C(u), \tag{16}
$$

where the set C is assumed to be closed and star-shaped with respect to a certain ball, $f \in X'$ and $A : X \to X'$ is assumed to be a pseudomonotone operator.

Problem (16) originates from the engineering literature [90], [111] and [137]. Indeed, Problem (16) constitutes a general mathematical form of the principle of virtual works for problems in mechanics whose material laws or boundary conditions result from nonconvex, generally nondifferentiable, potentials. Problem (16) can also be seen as a special case of a great class of inequality problems called hemivariational inequalities and which have been introduced by P.D. Panagiotopoulos [110] so as to formulate various problems in unilateral mechanics.

If in addition the set C is convex then Problem (16) reduces to the well-known variational inequality:

$$u \in C : \langle Au - f, v - u \rangle \geq 0, \forall v \in C. \tag{17}$$

If the set C is the whole space X then Problem (16) reduces to the usual variational equality

$$u \in X : \langle Au - f, v \rangle = 0, \forall v \in X. \tag{18}$$

Both of these two last cases have been the object of various studies.

The first mathematical theory applicable to Problem (16) when C is not convex is due to Z. Naniewicz [106]. The approach of Z. Naniewicz requires the coercivity of the operator A. Generalizations of this approach so as to be applicable to non-coercive problems have been proposed by D. Goeleven [72]. Several applications of Problem P can be found in D. Goeleven [72], D. Goeleven, G.E. Stavroulakis and P.D. Panagiotopoulos [77] and Z. Naniewicz and P.D. Panagiotopoulos [108].

Let us first recall some concepts which will be used later. An operator $T : X \to 2^{X'}$ is said to be pseudomonotone [36] if

(i) the set Tu is nonempty, bounded, closed and convex for any $u \in X$;

(ii) T is upper semicontinuous from each finite dimensional subspace F of X to X' equipped with the weak topology, i.e. to a given element $f \in F$ and a weak neighborhood V of $T(f)$ in X' there exists a neighborhood U of f in F such that $T(u) \subset F$ for all $u \in U$;

(iii) if $u_n \rightharpoonup u$ and if $z_n \in T(u_n)$ is such that

$$\limsup \langle z_n, u_n - u \rangle \leq 0,$$

then for each $v \in V$ there exists $z(v) \in T(u)$ such that

$$\liminf \langle z_n, u_n - v \rangle \geq \langle z(v), u - v \rangle.$$

An operator $A : X \to X'$ is said to have the S^+-property [122] if $u_n \rightharpoonup u$ and $\limsup \langle Au_n, u_n - u \rangle \leq 0$ implies that $u_n \to u$.

Let $C \subset X$ be a nonempty closed subset. We denote by

$$T_C(u) := \{k \in X : \forall u_n \in C, u_n \to u, \forall \lambda_n \downarrow 0,$$
$$\exists k_n \to k : u_n + \lambda_n . k_n \in C\},$$

the Clarke's tangent cone of C at u, by

$$N_C(u) := \{u^* \in X' : \langle u^*, k \rangle \leq 0, \forall k \in T_C(u)\},$$

the Clarke's normal cone to C at u, by

$$d_C(u) := \inf_{w \in C} \| u - w \|,$$

the distance function of C, by

$$d_C^0(u, v) := \limsup_{y \to u, t \downarrow 0} [d_C(y + tv) - d_C(y)]/t,$$

the generalized directional derivative of d_C at u in the direction v and by

$$\partial d_C(u) := \{w \in X', d_C^0(u, v) \geq \langle w, v \rangle, \forall v \in X\},$$

the Clarke's generalized gradient of d_C at u [48].

Let $B(u_0, \rho)$ be a closed ball in X with center u_o and radius $\rho > 0$. We say that C is star-shaped with respect to $B(u_o, \rho)$ [106] if

$$v \in C \Leftrightarrow \lambda v + (1 - \lambda)w \in C, \forall \lambda \in [0, 1], \forall w \in B(u_o, \rho).$$

We resume in the following Lemma the basic results concerning the function distance of a star-shaped set which have been proved by Z. Naniewicz.

Lemma 3.3.1. ([106]; Z. Naniewicz). Let X be a real reflexive Banach space, C a nonempty closed subset of X. If C is star-shaped with respect to $B(u_o, \rho)$ then

1) $d_C^0(u, u_o - u) \leq -d_C(u) - \rho, \forall u \notin C$

2) $d_C^0(u, u_o - u) = 0, \forall u \in C.$

The following result concerning the pseudomonotonicity property of the generalized Clarke's gradient is also due to Z. Naniewicz [105].

Lemma 3.3.2. ([105]; Z. Naniewicz). Let X be a real reflexive Banach space. Let $f_i : X \to \mathbb{R}$ be a finite collection of locally Lipschitzian convex functions defined on X. Define $f : X \to \mathbb{R}$ as

$$f(u) := \min\{f_i(u) : i = 1, ...N\}, u \in X.$$

Let $A : X \to X'$ be a maximal monotone operator with $D(A) = X$ and satisfying the S^+-property. Then $A + \partial f$ is pseudomonotone.

As a direct consequence of this Lemma, we get the following result.

Proposition 3.3.1. Let X be a real reflexive Banach space, $A : X \to X'$ a maximal monotone operator with $D(A) = X$ and satisfying the S^+-property. Let C be a subset of X which can be represented as the union of a finite collection of nonempty closed convex subsets $C_j(j = 1, ..., N)$ of X, i.e. $C = \cup_{j=1}^{N} C_j$. We assume that $int\{\cap_{j=1}^{N} C_j\} \neq \emptyset$. Then i) C is star-shaped with respect to a certain ball and ii) for each $\lambda \geq 0, A + \lambda \partial d_C$ is pseudomonotone.

Proof: i) trivial. ii) The distance function of C is expressed as a pointwise minimum of the Lipschitzian convex functions $d_i : X \to \mathbb{R}$, where d_i denotes the distance function of C_i, and the result follows from Lemma 3.3.2.

■

Let $u_0 \in X$ be given. We set

$$\Delta_{u_o} = \{x \in C \mid \langle Ax, x - u_o \rangle \leq \langle f, x - u_o \rangle\}$$

and we consider the recession set $R(\Delta_{u_o})$.

As in Section 3.1, we say that $R(\Delta_{u_o})$ is a-compact if the following property holds true. If $\{w_n; n \in \mathbb{N}\}$ is a sequence such that

$$w_n := u_n / \| u_n \| \rightharpoonup w,$$

with

$$u_n \in \Delta_{u_o}$$

and

$$\| u_n \| \to +\infty,$$

then there exists a subsequence satisfying $w_n \rightharpoonup w$.

Let us now give some properties of the recession set.

Proposition 3.3.2. Let u_o be given in X and f in X'. If

(i) A satisfies the S^+-property;

(ii) $\langle Ax, x \rangle \geq 0, \forall x \in X$;

(iii) A is weakly continuous, i.e. $x_n \rightharpoonup x \Rightarrow Ax_n \rightharpoonup Ax$;

(iv) A is positively homogeneous.

Then $R(\Delta_{u_o})$ is a-compact and

$$R(\Delta_{u_o}) \subset \{w \in C_\infty \backslash \{0\} \mid \langle Aw, w \rangle = 0\}.$$

Proof: Let $w \in R(\Delta_{u_o})$. There exists $u_n \in C$ such that $t_n := \| u_n \| \to +\infty, w_n := u_n/t_n \rightharpoonup w$, and

$$\langle Au_n, u_n - u_o \rangle \leq \langle f, u_n - u_o \rangle \tag{19}$$

Dividing (19) by t_n^2, we obtain

$$\langle Aw_n, w_n \rangle \leq \langle Aw_n, \frac{u_o}{t_n} \rangle + \langle \frac{f}{t_n}, w_n - \frac{u_o}{t_n} \rangle \tag{20}$$

and thus, by assumption (iii),

$$\limsup \langle Aw_n, w_n \rangle \leq 0. \tag{21}$$

53

We have

$$\limsup\langle Aw_n, w_n - w\rangle \leq \limsup\langle Aw_n, w_n\rangle + \limsup\langle Aw_n, -w\rangle,$$

so that, by assumption (ii) and (iii)

$$\limsup\langle Aw_n, w_n - w\rangle \leq \limsup\langle Aw_n, w_n\rangle.$$

This together with (21) imply that

$$\limsup\langle Aw_n, w_n - w\rangle \leq 0,$$

and thus, by assumption (i), the sequence w_n is strongly convergent to $w \in C_\infty$, which proves the a-compacity of $R(\Delta_{u_o})$. Since $\| w_n \| = 1$ and $w_n \to w$, we have $\| w \| = 1$. Then using (20) again, we get $\langle Aw, w\rangle \leq 0$. Therefore, using assumption (ii) again, we obtain

$$R(\Delta_{u_o}) \subset \{w \in C_\infty \backslash \{0\} \mid \langle Aw, w\rangle = 0\}.$$

$\blacksquare$

Example 3.3.1. Set $X := H^1(\Omega)$, where Ω is an open bounded subset of class C^1 in $\mathbb{R}^n (n \geq 1, n \in \mathbb{N})$. Let $A : X \to X'$ be the bounded linear operator defined by

$$\langle Au, v\rangle := \int_\Omega \nabla u.\nabla v dx, \forall u, v \in X.$$

It is easy to see that assumptions (ii)-(iv) are satisfied. It remains to prove that A satisfies the S^+-property. Indeed, let $\{w_n; n \in \mathbb{N}\}$ be a sequence such that $w_n \rightharpoonup w$ in $H^1(\Omega)$ (which implies that $w_n \to w$ in $L^2(\Omega)$ and $\nabla w_n \rightharpoonup \nabla w$ in $L^2(\Omega)$) and

$$\limsup \int_\Omega \nabla w_n.\nabla(w_n - w)dx \leq 0.$$

We get

$$\limsup \int_\Omega |\nabla w_n|^2 dx \leq \limsup \int_\Omega \nabla w_n.\nabla(w_n - w)dx$$
$$+ \limsup \int_\Omega \nabla w_n.\nabla w dx$$
$$\leq \int_\Omega |\nabla w|^2 dx.$$

Thus by using the weak lower semicontinuity of the map $x \to \langle Ax, x \rangle$, we obtain

$$\int_\Omega |\nabla w|^2 \, dx \leq \liminf \int_\Omega |\nabla w_n|^2 \, dx$$
$$\leq \limsup \int_\Omega |\nabla w_n|^2 \, dx \leq \int_\Omega |\nabla w|^2 \, dx.$$

Thus

$$\| \nabla w_n \|_{L^2} \to \| \nabla w \|_{L^2} .$$

Since $\nabla w_n \rightharpoonup \nabla w$ and the norm in $L^2(\Omega)$ is Kadec, we obtain

$$\nabla w_n \to \nabla w \text{ in } L^2(\Omega). \tag{22}$$

Using the fact that $w_n \to w$ in $L^2(\Omega)$ together with (22), we get

$$w_n \to w \text{ in } H^1(\Omega).$$

■

Proposition 3.3.3. Let u_o be given in X and f in X'. If

(i) $R(\Delta_{u_o})$ is a-compact;

(ii) there exists a nonempty subset W of X such that

$$R(\Delta_{u_o}) \subset W \backslash \{0\}$$

and

(iii) $r_{u_o,A}(w) > \langle f, w \rangle, \forall w \in W \backslash \{0\}$.

Then $R(\Delta_{u_o})$ is empty.

Proof. Suppose by contradiction that $R(\Delta_{u_o})$ is nonempty. Since $R(\Delta_{u_o}) \subset W \backslash \{0\}$ we obtain

$$r_{u_o,A}(w) > \langle f, w \rangle, \forall w \in R(\Delta_{u_o}).$$

55

We can also find a sequence $\{u_n \mid n \in I\!N\}$ such that $t_n := \|u_n\| \to \infty$, $w_n := u_n / \|u_n\| \to w$

and

$$\langle A(t_n w_n), t_n w_n - u_o \rangle \leq \langle f, t_n w_n - u_o \rangle \qquad (23)$$

Dividing (23) by t_n, we get

$$\langle A(t_n w_n) t_n^{-1}, t_n w_n - u_o \rangle \leq \langle f, w_n - \frac{u_o}{t_n} \rangle.$$

By assumption (i), $w_n \to w$ and thus

$$r_{u_o, A}(w) \leq \langle f, w \rangle,$$

which is a contradiction to assumption (iii).

∎

Proposition 3.3.4. Let X be a real reflexive Banach space such that $X \hookrightarrow L^2(\Omega)$ continuously (Ω denotes an open set in $I\!R^n$). Let u_o be given in X and f in X'. We assume that $f(x, u) : \Omega \times I\!R \to I\!R$ is measurable in x and continuous in u. Assume for a.e. $x \in \Omega$ and all $u \in I\!R$,

(f_1) $\mid f(x, u) \mid \leq a. \mid u \mid + b(x), a \geq 0, b \in L^2(\Omega)$
(f_2) $u.f(x, u) \geq -c(x). \mid u \mid - d(x), c \in L^2(\Omega), d \in L^1(\Omega)$.

Let $A_1 : X \to X'$ be an operator satisfying assumptions (i)-(iv) of Proposition 3.3.2 and let $A_2 : X \to X'$ be the operator defined by

$$\langle A_2 u, v \rangle := \int_\Omega f(x, u).v dx, \forall u, v \in X.$$

We set

$$A := A_1 + A_2.$$

Then $R(\Delta_{u_o})$ is a-compact and

$$R(\Delta_{u_o}) \subset \{w \in C_\infty \backslash \{0\} \mid \langle A_1 w, w \rangle = 0\}.$$

Proof: Let $w \in R(\Delta_{u_o})$. There exists $u_n \in X$ such that $t_n := \| u_n \| \to +\infty, w_n := u_n/t_n \rightharpoonup w$ and

$$\langle A_1 w_n, w_n \rangle + \langle A_2 u_n, w_n \rangle/t_n \leq \langle A_1 w_n, \frac{u_o}{t_n} \rangle$$

$$+ \langle A_2 u_n, u_0 \rangle/t_n^2 + \langle \frac{f}{t_n}, w_n - \frac{u_o}{t_n} \rangle$$

which means that

$$- \int_\Omega \frac{c(x)}{t_n^2} \mid u_n \mid dx - \int_\Omega \frac{d(x)}{t_n^2} dx + \langle A_1 w_n, w_n \rangle$$

$$\leq \langle A_1 w_n, \frac{u_o}{t_n} \rangle + a. \int_\Omega \frac{\mid u_n \mid}{t_n^2} \mid u_o \mid dx$$

$$+ \int_\Omega \frac{b(x)}{t_n^2} \mid u_0 \mid dx + \langle \frac{f}{t_n}, w_n - \frac{u_o}{t_n} \rangle.$$

The embedding $X \hookrightarrow L^2(\Omega)$ is continuous and there exists $C > 0$ such that

$$\| u \|_{L^2} \leq C \| u \|, \forall u \in X.$$

Then it is easy to see that

$$\langle A_1 w_n, w_n \rangle \leq \langle A_1 w_n, \frac{u_o}{t_n} \rangle + \langle \frac{f}{t_n}, w_n - \frac{u_o}{t_n} \rangle + \alpha/t_n + \beta/t_n^2,$$

where

$$\alpha = C(\| c \|_{L^2} + a \| u_o \|_{L^2})$$

and

$$\beta = \| d \|_{L^1} + \| b \|_{L^2} \| u_o \|_{L^2}.$$

Thus

$$\limsup \langle A_1 w_n, w_n \rangle \leq 0,$$

and we conclude as in Proposition 3.3.2.

∎

We now state the main Theorem of this Section.

Theorem 3.3.1. Suppose that the following assumptions are satisfied.

(H$_1$) X is a real reflexive Banach space and C is a nonempty closed subset of X which is star-shaped with respect to a ball $B(u_o, \rho), \rho > 0$;

(H$_2$) $A + \lambda \partial d_C$ is pseudomonotone for each $\lambda > 0$;

(H$_3$) A is bounded.

If $R(\Delta_{u_o}) = \emptyset$ then there exists $u \in C$ such that

$$\langle Au - f, v \rangle \geq 0, \forall v \in T_C(u).$$

Proof: Fix $n \in I\!N \backslash \{0\}$ and let

$$B_k := \{x \in X : \| x \| \leq k\},$$

where $k \in I\!N \backslash \{0\}$ is chosen great enough so that $u_o \in B_k$.

Let $j \geq k$ be given in $I\!N$. Since d_C is Lipschitz continuous, the operator $\partial d_C : X \to X'$ acts as a bounded operator, so that with assumptions (H$_1$)-(H$_3$) and by ([35]; Theorem 7.8) there exist

$$u_{n,j} \in B_j$$

and

$$z_{n,j} \in (A + n \partial d_C)(u_{n,j})$$

such that

$$\langle z_{n,j} - f, v - u_{n,j} \rangle \geq 0, \forall v \in B_j.$$

Thus we get

$$n d_C^0(u_{n,j}, v - u_{n,j}) + \langle Au_{n,j} - f, v - u_{n,j} \rangle \geq 0, \forall v \in B_j.$$

We claim that there exists $\theta = \theta(j) \in \mathbb{N}\backslash\{0\}$ such that $u_{\theta,j} \in C$. Indeed, suppose on the contrary that $u_{n,j} \notin C, \forall n \in \mathbb{N}\backslash\{0\}$. Then

$$\langle Au_{n,j} - f, u_o - u_{n,j}\rangle + nd^0_C(u_{n,j}, u_o - u_{n,j}) \geq 0,$$

which implies that

$$\langle Au_{n,j} - f, u_o - u_{n,j}\rangle \geq nd_C(u_{n,j}) + n\rho \geq n\rho.$$

Thus

$$\begin{aligned} n\rho &\leq \parallel f \parallel_* \parallel u_o - u_{n,j} \parallel + \parallel Au_{n,j} \parallel_* \parallel u_o - u_{n,j} \parallel \\ &\leq \sigma(j) \end{aligned}$$

with $\sigma(j) := \parallel f \parallel_* (j + \parallel u_o \parallel) + \parallel A \parallel j(j + \parallel u_o \parallel)$. We conclude that for each $n \in \mathbb{N}\backslash\{0\}$

$$n \leq \sigma(j),$$

which is a contradiction.

We prove that there exists $k' \in \mathbb{N}, k' \geq k$ such that $\parallel u_{\theta(k'),k'} \parallel < k'$. If not, $\parallel u_{\theta(i),i} \parallel = i$ for each $i \in \mathbb{N}, i \geq k$. On relabeling if necessary, the sequence defined by $w_i := w_{\theta(i),i} = u_{\theta(i),i}/i$ satisfies $w_i \rightharpoonup w, u_i := u_{\theta(i),i} \in C$ and

$$\langle Au_i - f, u_i - u_o\rangle \leq 0,$$

which means that $w \in R(\Delta_{u_o})$, a contradiction.

We have

$$\theta(k')d^0_C(u_{k'}, v - u_{k'}) + \langle Au_{k'} - f, v - u_{k'}\rangle \geq 0, \forall v \in B_{k'}. \tag{24}$$

Let $y \in X$, there exists $\varepsilon > 0$ such that

$$u_{k'} + \varepsilon(y - u_{k'}) \in B_{k'}.$$

59

If we put $v := u_{k'} + \varepsilon(y - u_{k'})$ in (24), we get

$$\theta(k')\varepsilon d_C^0(u_{k'}, y - u_{k'}) + \varepsilon\langle Au_{k'} - f, y - u_{k'}\rangle \geq 0.$$

Since $\varepsilon > 0$, y is arbitrarely chosen in X and $u_{k'} \in C$, we obtain

$$\theta(k')d_C^0(u_{k'}, y) + \langle Au_{k'} - f, y\rangle \geq 0, \forall y \in X. \qquad (25)$$

Since $u_{k'} \in C$ we have

$$y \in T_C(u_{k'}) \Leftrightarrow d_C^0(u_{k'}, y) = 0$$

and therefore inequality (25) entails that $u_{k'}$ is a solution of Problem (16).

∎

Corollary 3.3.1 and Corollary 3.3.3 stated below are for nonvariational problems what corollaries 3.1.3 and corollaries 3.1.4 are for the variational ones.

Corollary 3.3.1. Suppose that assumptions (H_1)-(H_3) are satisfied. If

(i) $R(\Delta_{u_o})$ is a-compact,

(ii) there exists a nonempty subset W of X such that
$R(\Delta_{u_o}) \subset W \backslash \{0\}$,

(iii) $\underline{r}_{u_o, A}(w) > \langle f, w \rangle, \forall w \in W \backslash \{0\}$,

then there exists $u \in C$ such that

$$\langle Au - f, v \rangle \geq 0, \forall v \in T_C(u).$$

Proof: We claim that $R(\Delta_{u_o}) = \emptyset$. If not then using (i) and (ii) we get a sequence $w_n := u_n / \| u_n \| \to w \in W \backslash \{0\}$, with

$$\| u_n \| \to +\infty$$

60

and

$$\langle Au_n, u_n - u_o \rangle \leq \langle f, u_n - u_o \rangle.$$

Thus

$$\underline{r}_{u_o,A}(w) \leq \langle f, w \rangle.$$

It is a contradiction with assumption (iii) and thus $R(\Delta_{u_o}) = \emptyset$. ∎

Corollary 3.3.2. Let X be a real reflexive Banach space and let $A : X \to X'$ be a bounded linear operator. We suppose that assumptions (H_1) and (H_2) are satisfied. If A is coercive, i.e. there exists $\alpha > 0$ such that

$$\langle Au, u \rangle \geq \alpha. \| u \|^2, \forall u \in X.$$

Then for each $f \in X'$, then there exists $u \in C$ such that

$$\langle Au - f, v \rangle \geq 0, \forall v \in T_C(u).$$

Proof: It is clear that

$$\langle Au, u - u_o \rangle / \| u \| \to +\infty \text{ as } \| u \| \to +\infty,$$

thus $R(\Delta_{u_o})$ is empty and we can conclude by application of Theorem 3.3.1. ∎

Corollary 3.3.3. Let X be a real Hilbert space and let $A : X \to X'$ be a bounded linear operator. We suppose that assumptions (H_1) and (H_2) are satisfied. If

(i) A is semicoercive, i.e. there exists $\alpha > 0$ such that

$$\langle Au, u \rangle \geq \alpha. \| Pu \|^2, \forall u \in X,$$

with $P = I - Q$, where I denotes the identity mapping and Q denotes the orthogonal projector of X onto $Ker(A + A^*)$ (A^* is the adjoint operator of A);

61

(ii) $\dim\{Ker(A + A^*)\} < +\infty$;

(iii) $u_o \in KerA$;

(iv) $\langle f, w \rangle < 0, \forall w \in C_\infty \cap Ker(A + A^*)\backslash\{0\}$.

Then there exists $u \in C$ such that

$$\langle Au - f, v \rangle \geq 0, \forall v \in T_C(u).$$

Proof: We will prove that all assumptions of Corollary 3.3.1 are satisfied. We claim that A satisfies the S^+-property. Indeed, let $\{u_n; n \in I\!N\}$ be a sequence such that

$$u_n \rightharpoonup u \text{ in } X,$$

and

$$\limsup \langle Au_n, u_n - u \rangle \leq 0.$$

We have

$$
\begin{aligned}
\alpha. \limsup \| Pu_n - Pu \|^2 &\leq \limsup \langle Au_n - Au, u_n - u \rangle \\
&\leq \limsup \langle Au_n, u_n - u \rangle + \limsup \langle Au, u - u_n \rangle \\
&\leq 0,
\end{aligned}
$$

and thus $Pu_n \to Pu$. Moreover, Q is bounded linear and thus weakly continuous. Therefore $Qu_n \to Qu$ since $\dim\{Ker(A + A^*)\} < +\infty$. Thus

$$u_n = Pu_n + Qu_n \to Pu + Qu = u.$$

It is also clear that all other assumptions required by Proposition 3.3.1 are also satisfied. Thus $R(\Delta_{u_o})$ is a-compact and $R(\Delta_{u_o}) \subset W\backslash\{0\}$, where

$$W = C_\infty \cap \{x \in X \mid \langle Ax, x \rangle = 0\} = C_\infty \cap Ker(A + A^*).$$

We have

$$\langle A(tx), tx - u_o \rangle / t \geq -\langle Ax, u_o \rangle, \forall x \in X.$$

Thus

$$\underline{r}_{u_o,A}(w) \geq -\langle Aw, u_o \rangle = 0, \forall w \in W \backslash \{0\}, \tag{26}$$

so that assumption (iv) together with (26) imply assumption (iii) in Corollary 3.3.1.

∎

Remarks 3.3.1. i) If A is semicoercive, C is a subset of X which can be represented as the union of a finite collection of nonempty closed convex subsets $C_j (j = 1, ..., N)$ of X, i.e. $C = \cup_{j=1}^{N} C_j$ and if $u_o \in int\{\cap_{j=1}^{N} C_j\}$, then assumptions (H$_1$) and (H$_2$) are satisfied. Indeed, we known that A is maximal monotone and satisfies the S^+-property so that we may conclude by using Proposition 3.3.1. ii) if A is symmetric, then $W = C_\infty \cap (KerA) \backslash \{0\}$ and $\underline{r}_{u_o,A}(w) \geq 0, \forall w \in W$, so that assumption (iii) on u_o is not necessary. iii) If the set C is additionally supposed to be convex then the recession approach can be refined in various directions. It is out of the scope of this work to consider this case which need further investigations. For the study of noncoercive variational inequalities invoking nonvariational operators, we refer to S. Adly, D. Goeleven and M. Théra [3],D.D. Ang, K. Schmitt and L.K. Vy [6], D. Goeleven [67], M. Schatzman [125] and F. Tomarelli [132]. For the study of other kinds of hemivariational inequalities invoking nonvariational operators, we refer to S. Adly, D. Goeleven and M. Théra [4], Z. Naniewicz and P.D. Panagiotopoulos [108] and D. Goeleven and M. Théra [79]. iv) Corollaries 3.1.3, 3.1.4, 3.3.1 and 3.3.3 constitute basic results which will be used several times in the following applications (chapter 4).

4 APPLICATIONS

4.1 An elliptic problem with unilateral boundary conditions

To begin this chapter on applications, we choose a semicoercive problem of relatively simple but illustrative character.

Let $\Omega \subset I\!\!R^N$ be a bounded open regular subset (i.e. with a C^1-boundary Γ and Ω located on one side of Ω). We consider the elliptic operator

$$Lu = \sum_{i,j=1}^{N} \frac{\partial}{\partial x_j}(a_{ij}\frac{\partial u}{\partial x_i}) + c.u$$

with

$$a_{ij}, c \in L^{\infty}(\Omega), \tag{27}$$

$$a_{ij} = a_{ji}, c \geq 0. \tag{28}$$

$$\sum_{i,j=1}^{N} a_{ij}\zeta_i\zeta_j \geq \alpha\,|\,\zeta\,|^2 \text{ a.e. on } \Omega, \forall \zeta \in I\!\!R^N (\alpha > 0). \tag{29}$$

Let $a : H^1(\Omega) \times H^1(\Omega) \to I\!\!R$ be the bilinear, symmetric and continuous form defined by

$$a(u,v) = \int_{\Omega} \sum_{i,j=1}^{N} a_{ij}\frac{\partial u}{\partial x_i}\frac{\partial v}{\partial x_j} + c.uvdx.$$

Let $A : H^1(\Omega) \to H^1(\Omega)'$ be the corresponding bounded, linear and self-adjoint operator defined by

$$\langle Au, v \rangle = a(u,v), \forall u,v \in H^1(\Omega).$$

Here

$$Ker\{A\} = \begin{cases} I\!\!R & \text{if } c = 0, \\ \\ 0 & \text{if } c > 0. \end{cases}$$

64

Let $\Phi : L^2(\Gamma) \rightarrow (-\infty, +\infty]$ be a proper convex and lower semicontinuous functional on $L^2(\Gamma)$. The graph $\partial\Phi$ is maximal monotone in $L^2(\Gamma) \times L^2(\Gamma)$. By the Hahn-Banach Theorem, there exists $\alpha_1 \geq 0$ and $\alpha_2 \in I\!R$ such that

$$\Phi(v) \geq -\alpha_1 \parallel v \parallel_{\Gamma,0,2} -\alpha_2, \forall v \in L^2(\Gamma).$$

We denote by Φ^* the restriction of Φ on $H^{\frac{1}{2}}(\Gamma)$ and we assume that $\mathrm{dom}\{\Phi^*\} \neq \emptyset$. Then the graph $\partial\Phi^*$ is maximal monotone in $H^{\frac{1}{2}}(\Gamma) \times H^{-\frac{1}{2}}(\Gamma)$.

For $f \in L^2(\Omega)$ be given, we consider the problem of finding a minimizer for the functional

$$\psi(u) := \tfrac{1}{2}a(u,u) + \Phi(\gamma_o(u)) - \int_\Omega f.u dx \tag{30}$$

on $H^1(\Omega)$. Here $\gamma_o : H^1(\Omega) \rightarrow H^{\frac{1}{2}}(\Gamma)$ denotes the linear and continuous trace operator [122]. Note that the composite function $u \in H^1(\Omega) \rightarrow \Phi[\gamma_o](u) := \Phi(\gamma_o(u))$ is convex, lower semicontinuous and $(\Phi[\gamma_o])_\infty(u) = \Phi_\infty(\gamma_o(u))$.

We denote by $\Im$ the solutions set of problem (30), i.e.

$$\Im := \mathrm{argmin}\{\psi(x) \mid x \in H^1(\Omega)\}.$$

Theorem 4.1.1. If

$$\int_\Omega f.e dx < \Phi_\infty(\gamma_o(e)), \forall e \in Ker\{A\}\backslash\{0\}. \tag{31}$$

Then there exists $u \in \Im$ such that

$$(1) \quad a(u, v - u) + \Phi(\gamma_o(v)) - \Phi(\gamma_o(u)) \geq \int_\Omega f(v-u)dx,$$
$$\forall v \in H^1(\Omega).$$

Moreover u satisfies the following conditions:

$$(2) \quad Lu = f \text{ in } \mathcal{D}'(\Omega)$$

and

$$(3) \quad -\sum_{i,j=1}^{N} a_{ij}\frac{\partial u}{\partial x_i}\cos(n, x_j) \in \partial\Phi^*(\gamma_o(u)).$$

(Here n denotes the outward normal to Ω and $\mathcal{D}'(\Omega)$ is the space of distributions).

Proof: We claim that $R(\Delta(\varepsilon_n)) \subset Ker(A)$. Indeed, if $w \in R(\Delta(\varepsilon_n))$ then by Proposition 3.1.1 (4), there exists a sequence $\{u_n; n \in \mathbb{N}\}$ such that $\| u_n \|_{1,2} \to +\infty, w_n := u_n \cdot \| u_n \|_{1,2}^{-1} \rightharpoonup w$ and

$$\limsup \psi(u_n) / \| u_n \|_{1,2}^2 \leq 0.$$

Therefore

$$
\begin{aligned}
\liminf \langle Aw_n, w_n \rangle \quad &+ \quad \liminf\{-\alpha_1 \| u_n \|_{1,2}^{-2} \| \gamma_o(u_n) \|_{\Gamma,0,2} -\alpha_2 \| u_n \|_{1,2}^{-2}\} \\
&\leq \quad \limsup\{[\langle Au_n, u_n \rangle + \Phi(\gamma_o(u_n))] \cdot \| u_n \|_{1,2}^{-2}\} \\
&\leq \quad 0,
\end{aligned}
$$

and thus $w \in Ker(A)$. Moreover $w_n \to w$ in $L^2(\Omega), \nabla w_n \rightharpoonup \nabla w = 0$ and $\| \nabla w_n \|_{0,2} \to 0$. Thus $w_n \to w$ in $H^1(\Omega)$ and $R(\Delta(\varepsilon_n))$ is a-compact. We may conclude by using Corollary 3.1.3 since assumption (31) implies that

$$\psi_\infty(e) > 0, \forall e \in Ker(A)\backslash\{0\}.$$

Moreover, if $u \in \mathfrak{S}$ then $u \in H^1(\Omega)$ and satisfies the following variational inequality [56]

$$a(u, v - u) + \Phi(\gamma_o(v)) - \Phi(\gamma_o(u)) \geq \int_\Omega f(v - u)dx, \forall v \in H^1(\Omega).$$

Therefore, by a result of H. Brézis ([31]; Theorem 1.7), conditions (2) and (3) are satisfied.

∎

Remarks 4.1.1.

i) If $c > 0$ then condition (31) is always satisfied since $Ker\{A\}\backslash\{0\} = \emptyset$.

ii) If $c = 0$, then using Propositions 2.1.5, we see that (31) is equivalent to

$$f \in \text{int}\{I\!\!R^\perp + conv \ R(\partial(\Phi[\gamma_o]))\}.$$

Here

$$I\!\!R^\perp = \{u \in H^1(\Omega) \mid \int_\Omega u(x)dx = 0\}.$$

iii) Theorem 4.1.1 can also be obtained by application of Corollary 3.1.4.

iv) If $\Im \neq \emptyset$ then

$$\int_\Omega f.edx \leq \Phi_\infty(\gamma_o(e)), \forall e \in Ker\{A\}.$$

v) If u_1 and $u_2 \in \Im$ then there exists $c \in I\!\!R$ such that $u_1 = u_2 + c$.

4.2 Equilibrium of masonry structures

In this Section we consider a model proposed by M. Giaquinta and E. Giusti [65] for the description of masonry-like problem.

Let Ω be a star-shaped open and bounded subset of $I\!\!R^2$, with Lipschitz-continuous boundary Γ.

We denote by $BD(\Omega)$ the space

$$BD(\Omega) = \{u \in L^1(\Omega, I\!\!R^2) \mid \varepsilon(u) \in \mathbf{M}(\Omega)\}$$

where $\mathbf{M}(\Omega)$ is the space of (2x2 matrix-valued) measures on Ω with bounded total variation, i.e. $\int_\Omega \mid \varepsilon(u) \mid < +\infty$, with

$$\int_\Omega \mid \varepsilon(u) \mid := \sup\{\tfrac{1}{2} \int_\Omega [u_i v_{ij,j} + u_j v_{ij,i}]dx \mid v \in C_o^\infty(\Omega, I\!\!R^{2\times2}), \ v : v \leq 1\}.$$

It is a Banach space with the norm

$$\| u \|_{BD(\Omega)} = \| u \|_{0,1} + \int_\Omega \mid \varepsilon(u) \mid.$$

We shall say that $u_n \xrightarrow{\tau} u$ if [130]

$$u_n \to u \text{ in } L^1(\Omega)$$

and

$$\varepsilon(u_n) \to \varepsilon(u) \text{ weakly (in the sense of measures)}.$$

Problems of static equilibrium for a class of elastic materials (masonry like-materials) in wich the stress is constrained to be negative semi-definite lead to the following model for the stored energy functional [65]

$$\int_\Omega |P_K(\varepsilon(u))|^2$$

where P_K is the orthogonal projection onto a subcone K of the cone $\mathbf{S}^2_-$ of negative semi-definite symmetric matrices of order 2, defined as follows

$$K := \{A^{-1}\eta \mid \eta \in \mathbf{S}^2_-\}$$

where $A : \mathbf{S}^2 \to \mathbf{S}^2$ ($\mathbf{S}^2$ denotes the space of symmetric 2x2 matrices) is a symmetric positive definite linear operator.

The integral is in the sense of convex function of a measure. If we decompose $\varepsilon(u)$ into its absolutely continuous and singular part with respect to the 2-dimensional Lebesgue measure, i.e.

$$\varepsilon(u) = \varepsilon^a(u) + \varepsilon^s(u),$$

and if we set

$$E(u) := |P_K(\varepsilon(u))|^2,$$

then we have

$$\int_\Omega E(u) = \int_\Omega E(\varepsilon^a(u))dx + \int_\Omega E^\infty(\frac{d\varepsilon^s}{d|\varepsilon^s|})(u)\,|\varepsilon^s|(u)$$

68

where $| \varepsilon |$ is the absolute variation measure associated to ε, i.e. for $B \subset \Omega$,

$$| \varepsilon | (B) = \sup\{\sum_{i=1}^{n_0} \sqrt[3]{\varepsilon(B)} : \varepsilon(B) \mid \cup_{i=1}^{n_0} B_i \subset B, B_i \cap B_j = \emptyset \text{ if } i \neq j\}.$$

and

$$\frac{d\varepsilon^s}{d \mid \varepsilon^s \mid}(u) = \lim_{\rho \to 0^+} \frac{\varepsilon^s(B_\rho(u))}{\mid \varepsilon^s \mid (B_\rho(u))}$$

where $B_\rho(u) := \{y \in I\!R^2 \mid \| u - y \| < \rho\}$. For a complete description of this point, we refer to [9], [37], [65] and [130].

Let $f \in L^2(\Omega, I\!R^2)$ and $t \in L^\infty(\Omega, I\!R^2)$ denote respectively a body-force and a surface traction acting on the body Ω. The total external work corresponding to a virtual displacement u is then given by

$$- \int_\Omega f.u dx - \int_\Gamma t.\gamma(u) ds.$$

Here $\gamma : BD(\Omega) \to L^1(\Gamma, I\!R^2)$ denotes the linear trace operator. The existence of this trace operator follows from a result of M. Giaquinta and E. Giusti ([65]; Theorem 5.5). We recall also that for a Borel set $\Omega \subset I\!R^N$, the integral $\int_\Omega dx$ is defined by means of the $n-$dimensional Lebesgue measure μ while the integral $\int_\Gamma ds$ is defined by means of the Hausdorff measure H_{n-1} of dimension $n-1$.

We are looking for the minimizers of the total energy functional

$$\psi(u) = \int_\Omega \mid P_K(\varepsilon(u)) \mid^2 - \int_\Omega f.u dx - \int_\Gamma t.\gamma(u) ds$$

on the space $BD(\Omega)$.

We say that the load (f,t) satisfies the safe load condition [65] if there exists a tensor field $H \in L^\infty(\Omega, \mathbf{S}^2)$ such that

$$H + \beta I \in K \tag{32}$$

for some $\beta > 0$, and

$$\int_\Omega H : \varepsilon(\theta) dx = \int_\Omega f.\theta dx + \int_\Gamma t.\gamma(\theta) ds, \forall \theta \in H^1(\Omega, I\!R^2). \tag{33}$$

69

Lemma 4.2.1. Suppose that (f, t) satisfies conditions (32) and (33). Then the functional $\psi : BD(\Omega) \to I\!R$ is τ-lower semicontinuous.

Proof: We prove that $epi(\psi)$ is τ-closed. Let $\{u_n; n \in I\!N\}$ and $\{t_n; n \in I\!N\}$ be sequences such that $u_n \xrightarrow{\tau} u, t_n \to t$ and $\psi(u_n) \leq t_n$. If $t = +\infty$ then it is clear that $\psi(u) \leq t$. If $t \in I\!R$ then $\limsup \psi(u_n) < +\infty$. Conditions (32) and (33) imply the existence of $c_o > 0$ and $c_1 \in I\!R$ such that ([65]; Theorem 6.6)

$$\psi(v) \geq c_o \int_\Omega \mid \varepsilon(v) \mid -c_1, \forall v \in BD(\Omega). \tag{34}$$

Then by using Sobolev embedding Theorem [130] and by considering a subsequence, we can assume that

$$u_n \rightharpoonup u \text{ in } L^2(\Omega),$$

$$u_n \to u \text{ in } L^1(\Omega)$$

and

$$\int_\Omega \mid \varepsilon(u_n) \mid \leq c$$

Then ([65]; Proposition 8.3) $\varepsilon(u_n) \rightharpoonup \varepsilon(u)$ weakly in measures and we can use a result of M. Giaquinta and E. Giusti ([65]; Theorem 6.7) in order to assert that

$$\psi(u) \leq \liminf \psi(u_n) \leq t.$$

■

Theorem 4.2.1. Suppose that (f, t) satisfies conditions (32) and (33). Then the functional $\psi : BD(\Omega) \to I\!R$ attains its minimum in the space $BD(\Omega)$.

Proof: Let $w \in R(\Delta(\varepsilon_n))$ be given. By Proposition 3.1.1, there exists a sequence $\{u_n; n \in I\!N\}$ such that

$$\parallel u_n \parallel \to +\infty,$$

70

$$w_n := u_n. \parallel u_n \parallel^{-1} \xrightarrow{\tau} w$$

and

$$\limsup \psi(u_n)/ \parallel u_n \parallel^p \le 0, \forall p > 0. \tag{35}$$

Inequality (34) together with (35) ($p = 1$) imply that

$$
\begin{aligned}
c_o \int_\Omega \mid \varepsilon(w) \mid \ &\le \ \liminf c_o \int_\Omega \mid \varepsilon(w_n) \mid \\
&\le \ \limsup c_o \int_\Omega \mid \varepsilon(w_n) \mid \\
&\le \ \limsup \{ c_o \int_\Omega \mid \varepsilon(w_n) \mid -c_1 \parallel u_n \parallel^{-1} \} + \limsup c_1 \parallel u_n \parallel^{-1} \\
&\le \ 0.
\end{aligned}
$$

Thus $\int_\Omega \mid \varepsilon(w_n) \mid \longmapsto \int_\Omega \mid \varepsilon(w) \mid$ and $w \in \mathbf{RBM} := \{\alpha \wedge x + \beta; \alpha, \beta \in I\!\!R^2\}$.

We have

$$\parallel w_n - w \parallel_{BD(\Omega)} = \parallel w_n - w \parallel_{0,1} + \int_\Omega \mid \varepsilon(w_n) \mid$$

since $\varepsilon(w) = 0$. Therefore $w_n \to w$ and $R(\Delta(\varepsilon_n))$ is $a(\tau)$-compact.

Moreover, if $w \in \mathbf{RBM}$ then

$$\psi(u - w) = \psi(u) + \int_\Omega f.w dx + \int_\Gamma t.\gamma(w)ds.$$

Condition (33) implies that

$$\int_\Omega f.w dx + \int_\Gamma t.\gamma(w)ds = 0$$

and thus $R(\Delta(\varepsilon_n)) \subset D_1$. The existence of a minimizer follows from Corollary 3.1.2.

∎

Remark 4.2.1

1) The first proof of Theorem 4.2.1 is due to M. Giaquinta and E. Giusti [65]. We have here recovered this result by using various properties proved by these authors and our general approach based on the recession analysis.

ii) For more details concerning the study of masonry structures, we refer the reader to [9], [10] and [65].

4.3 Global minimization of τ-lower semicontinuous functionals

The aim of this Section is to recover the celebrated result of C. Baiocchi, G. Buttazzo, F. Gastaldi and F. Tomarelli [21].

Let $\psi : X \to I\!R \cup \{+\infty\}$ be a proper τ-lower semicontinuous functional defined on a real Banach space X and satisfying (h_1) (see Section 3.1). To prove the existence of a global minimum for ψ, C. Baiocchi, G. Buttazzo, F. Gastaldi and F. Tomarelli [21] have introduced the following compatibility and compactness conditions.

Theorem 4.3.1. Suppose that the following compatibility and compactness conditions are satisfied.

COMPATIBILITY CONDITION

(i) $\tau - \psi_\infty(x) \geq 0, \forall x \in X,$

(ii) $\forall z \in Ker\{\tau - \psi_\infty(x)\}, \exists \mu > 0$ such that

$$\psi(x - \mu z) \leq \psi(x), \forall x \in X.$$

COMPACTNESS CONDITION

If $\{w_n; n \in I\!N\}$ and $\{\lambda_n; n \in I\!N\}$ are sequences such that $w_n \overset{\tau}{\to} w, \lambda_n \to +\infty$ and $\psi(\lambda_n w_n)$ is bounded from above, then $w_n \to w$.

Then

$$\operatorname{argmin}\{\psi(x) \mid x \in X\} \neq \emptyset.$$

We refind this result as a consequence of Corollary 3.1.2 and Remark 3.1.2 vi) or vii). Indeed, by Proposition 3.1.1, if $w \in R(\Delta(\varepsilon_n))$ then $w \in K_o(\tau - \psi_\infty)$. This together with assumption (i) imply that $w \in Ker\{\tau - \psi_\infty\}$. Therefore, by using assumption (ii), we get

$$R(\Delta(\varepsilon_n)) \subset Ker\{\tau - \psi_\infty\}$$

and for all $w \in R(\Delta(\varepsilon_n))$, there exists $\mu(w) > 0$ such that

$$w \in \{y \in X \mid \psi(u - \mu(w)y) \leq \psi(u), \forall u \in X\} = D_{\mu(w)}.$$

Moreover, by Proposition 3.1.1 (5), the sequence $\psi(\|u_n\|.w_n)$ is bounded from above so that the compactness condition stated above entails the $a(\tau)$-compacity of $R(\Delta(\varepsilon_n))$.

Remarks 4.3.1.

i) If $\tau - \psi_\infty(x) > 0, \forall x \in X \backslash \{0\}$ then the compatibility conditions (i) and (ii) are satisfied.

ii) If ψ is convex and τ-lower semicontinuous and if $Ker\{\psi_\infty\}$ is a subspace of X then the compatibility condition (ii) is satisfied [21]. More precisely, we have

$$Ker\{\psi_\infty\} \subset D_1.$$

4.4 Minimization of separable functionals in Hilbert space

Let X be a real Hilbert space. Let $X_1 \subset X$ be a closed vector subspace and $X_o = X_1^\perp$. Let $\psi : X \to I\!\!R \cup \{+\infty\}$ be a proper and weakly lower semicontinuous functional. For $u \in X$, we write $u = u_o + u_1$ with $u_o \in X_o$ and $u_1 \in X_1$. We assume that ψ can be split as the sum of two functionals $F, G : X \to I\!\!R \cup \{+\infty\}$, i.e. $\psi = F + G$, such that

(i) F is X_1 coercive (see Section 2.2), i.e.

$$F(x) \geq \alpha(\|x_1\|) \|x_1\|, \forall x \in X$$

73

with $\alpha(t) \to +\infty$ as $t \to +\infty$. Moreover, we assume that

$$\alpha(t) \geq 0, \forall t \geq 0.$$

(ii) there exist two functions $G_o : X_o \to I\!R \cup \{+\infty\}$ and $G_1 : X_1 \to I\!R$ such that:

$(ii - a)$ $G(x) \geq G_o(x_o) + G_1(x_1), \forall x \in X$;

$(ii - b)$ $| G_1(x) | \leq k(1 + \| x \|^\sigma), \forall x \in X_1 \ (k > 0, 0 < \sigma < +\infty)$;

$(ii - c)$ $G(x) \geq -\alpha_1 \| x_1 \|^\beta - \alpha_2, \forall x \in X \ (\alpha_1 \geq 0, \alpha_2 \in I\!R, 0 \leq \beta \leq 1)$.

Remark 4.4.1. Assumptions (ii-a) and (ii-c) are satisfied if $G : X \to I\!R$ is convex, lower semicontinuous and coercive on X_o. Indeed, let $a_o \in \text{argmin}\{G_o(x) \mid x \in X_o\}$. Then by a result of J. Mawhin [98] based on the Hann- Banach Theorem, there exists $a_1 \in X_1$ such that $a_1 \in \partial G(a_o)$. Therefore

$$
\begin{aligned}
G(x) &\geq G(a_o) + \langle a_1, x - a_o \rangle \\
&= G(a_o) + \langle a_1, x_1 \rangle \\
&\geq G(a_o) - \| a_1 \| \| x_1 \| .
\end{aligned}
$$

It is clear that assumption (ii-a) is satisfied since

$$G(x) \geq 2G(\tfrac{x_0}{2}) - G(-x_1).$$

In this Section, a Theorem of J. Mawhin [98] is reviewed for nonconvex functionals. We begin by recalling the result of J. Mawhin.

Theorem 4.4.1. ([98]; J. Mawhin) Suppose that assumption (i) is satisfied. If $G : X \to I\!R$ is a lower semicontinuous and convex function such that

(iii) $G(z) \to +\infty$ as $z \in X_o, \| z \| \to +\infty.$

Then

$$\text{argmin}\{\psi(x) \mid x \in X\} \neq \emptyset.$$

In this Section, we show that by using the recession analysis, we are able to obtain a similar result for functionals which are not necessarily convex.

For $c > 0$ be given, we set $\Delta(c) := \{x \in X_o \mid \psi(x) \leq c\}$ and as a rule, we define the recession set

$$R(\Delta(c)) := \{y \in X_o \mid \exists e_n \in \Delta(c), \| e_n \| \to +\infty$$

$$\text{and } y_n := e_n / \| e_n \| \to y\}.$$

We are now in position to refine Proposition 3.1.1.

Proposition 4.4.1. If $w \in R(\Delta(\varepsilon_n))$ then there exists a sequence $\{e_n; n \in I\!N\}$ such that

$$(1) \qquad e_n \in X_o;$$

$$(2) \qquad \| e_n \| \to +\infty;$$

$$(3) \qquad y_n := e_n / \| e_n \| \rightharpoonup y \in X_o;$$

$$(4) \qquad \limsup G_o(e_n) / \| e_n \|^p \leq 0, \forall p > 0;$$

$$(5) \qquad \limsup G_o(e_n) < +\infty;$$

and

$$(6) \qquad w - \{G_o\}_\infty(y) \leq 0.$$

Moreover, if $R(\Delta(c))$ is a-compact (for any $c > O$) then there exists a sequence $\{e_n; n \in I\!N\}$ satisfying (1), (2), (4), (5)

$$(3') \qquad y_n \to y \in X_o \backslash \{0\}$$

and

$$(6') \qquad \{G_o\}_\infty(y) \leq 0.$$

75

Proof: Let $v_o \in \text{dom}(\psi)$ be given. As in Proposition 3.1.1, we prove that there exists a sequence $\{u_n; n \in I\!N\}$ such that $\| u_n \| \to +\infty$ and

$$\psi(u_n) \leq \psi(v_o) + \varepsilon_n \| v_o \| . \tag{36}$$

Inequality (36) implies that

$$\liminf \alpha(\| u_{n,1} \|) \| u_{n,1} \| + \liminf(-\alpha_1 \| u_{n,1} \|^\beta - \alpha_2) < +\infty.$$

Therefore the sequence $u_{n,1}$ is bounded and thus $\| u_{n,o} \| \to +\infty$.

We set $e_n := u_{n,o}$ and by considering a subsequence, we can assume that $y_n := e_n/ \| e_n \| \rightharpoonup y$. Inequality (36) implies that

$$G_o(e_n) + G_1(u_{n,1}) \leq \psi(v_o) + \varepsilon_n \| v_o \| .$$

Thus

$$G_o(e_n) \leq k(1+ \| u_{n,1} \|^\sigma) + \psi(v_o) + \varepsilon_n \| v_o \|,$$

so that for n great enough, we may assume that

$$G_o(e_n) \leq C_o$$

where C_o is some positive constant. Therefore

$$w - \{G_o\}_\infty(y) \leq 0.$$

The other assertions of Proposition 4.4.1 can be proved easily as in Proposition 3.1.1.

∎

Theorem 4.4.2. Suppose that assumptions (i) and (ii) are satisfied. Moreover, we assume that

(iii)

$$G_o(z) \to +\infty \text{ as } z \in X_o, \| z \| \to +\infty.$$

76

Then

$$\operatorname{argmin}\{\psi(x) \mid x \in X\} \neq \emptyset.$$

Proof: We claim that $R(\Delta(\varepsilon_n))$ is empty. If not, then (iii) will be in contradiction with Proposition 4.4.1 (5).

∎

Example 4.4.1. Let $A \in \mathbb{R}^{2n}$ be a symmetric positive semidefinite matrix of order n. We set $X_o := Ker(A)$, $X_1 := R(A)$, $F(x) := x^T A x$ and $G(x) := \| P_{Ker(A)} x \|^5 - \| x - P_{Ker(A)} x \|$ (here $P_{Ker(A)}$ denotes the orthogonal projector of $\mathbb{R}^n$ onto $Ker(A)$). It is easy to see that all assumptions of Theorem 4.4.2 are satisfied.

Theorem 4.4.3. Suppose that assumptions (i) and (ii) are satisfied. Moreover, we assume that

(iii) $R(\Delta(c))$ is $a-$compact (for any $c > 0$);

(iv) $\{G_o\}_\infty(z) > 0, \forall z \in X_o, z \neq 0.$

Then

$$\operatorname{argmin}\{\psi(x) \mid x \in X\} \neq \emptyset.$$

Proof: We claim that $R(\Delta(\varepsilon_n))$ is empty. If not, then (iv) will be in contradiction with Proposition 4.4.1 (6').

∎

4.5 Noncoercive problems in nonlinear elasticity

Let Ω be a bounded open connected subset of $\mathbb{R}^3$ with a boundary Γ that is Lipschitz continuous. The set $\overline{\Omega}$ is the reference configuration occupied by a homogeneous hyperelastic body in the absence of any applied force.

The corresponding mixed displacement-traction problem consists in the following system (see [22], [44] and [45]).

$$-\mathrm{div}\,\mathbf{T}(\nabla u) = f \text{ in } \Omega; \tag{37}$$

$$\mathbf{T}(\nabla u)n = g \text{ on } \Gamma_1, \tag{38}$$

where the mapping $\mathbf{T}$ (first Piola-Kirchhoff stress tensor) is defined by

$$\mathbf{T}(F) = \frac{\partial W}{\partial F}(F) \in M^3,$$

where M^3 denotes the set of all matrices of order 3. We require that the admissible deformations $u : \overline{\Omega} \to \mathbb{R}^3$ are orientation-preserving and locally invertible, i.e.

$$\det(\nabla u) > 0. \tag{39}$$

A way (at least formally [44]) to solve this system consists in finding the stationary points of the total energy

$$\psi(u) = \int_\Omega W(\nabla u)dx - \int_\Omega f.udx - \int_{\Gamma_1} g.uds.$$

Let Γ_1 and Γ_2 be disjoint (relatively) open subsets of Γ, with $H_2(\Gamma_2) > 0$ and $H_2(\Gamma \backslash (\Gamma_1 \cup \Gamma_2)) = 0$.

We assume that the energy function $W : M_+^3 \to \mathbb{R}$ is polyconvex, i.e. there exists a convex function $\mathbf{W} : M^3 \times M^3 \times]0, +\infty[\to \mathbb{R}$ such that (Here $M_+^3 = \{F \in M^3 \mid \det F > 0\}$)

$$W(F) = \mathbf{W}(F, \mathrm{adj}(F), \det(F)), \forall F \in M_+^3. \tag{40}$$

We suppose the following behavior as $\det(F) \to 0^+$.

If $F_n \to F$ in $M^3, H_n \to H$ in M^3 and $\delta_n \to 0$

then $W(F_n, H_n, \delta_n) \to +\infty.$ \tag{41}

Condition (41) means that it is energetically impossible to compress part of the body to zero volume.

Moreover, we assume the existence of $a \in \mathbb{R}, b > 0, p \geq 2, q \geq p/(p-1), r > 1$ such that

$$\mathbf{W}(F, H, \delta) \geq a + b(\mid F \mid^p + \mid H \mid^q + \delta^r). \tag{42}$$

(Here $\mid F \mid := (F : F)^{\frac{1}{2}}$). Finally, we assume that $f \in L^{p'}(\Omega; \mathbb{R}^3)$ and $g \in L^{p'}(\Gamma; \mathbb{R}^3)$ $(p'^{-1} + p^{-1} = 1)$.

Let Q be an open subset of $\mathbb{R}^3$. We suppose that the body is subjected to the following conditions.

Unilateral boundary conditions

$$u \notin Q \text{ for } x \in \Gamma_2; \tag{43}$$

$$\mathbf{T}(\nabla u)n = 0 \text{ if } u \notin \overline{Q} \tag{44}$$

$$\mathbf{T}(\nabla u)n = \alpha n', \alpha \leq 0 \text{ if } u \in \partial Q \tag{45}$$

where n denotes the unit normal vector along the surface Γ at the point x and where n' denotes the unit normal vector along the surface $u(\Gamma)$ at the point $u(x)$. Relations (43) and (44) mean that at those points where the deformed boundary $u(\Gamma_2)$ and the obstacle Q have a constant tangent plane, the Cauchy stress vector $T'n'$ is directed along the outer normal to the obstacle.

Locking constraint

$$L(\nabla u) \leq 0 \text{ in } \Omega; \tag{46}$$

where $L : M_+^3 \to \mathbb{R}$ is the locking function assumed to be polyconvex.

We extend the domain of Definition of $\mathbf{W}$ by setting $W(F, H, \delta) = +\infty$ if $\delta \leq 0$ and we note that $\mathbf{W}(F, H, \delta)$ is convex and continuous since (41) holds. We set

$$\Phi(u) := \int_\Omega W(\nabla u)dx.$$

Then problem (37)-(39), (43)- (46) leads us (at least formally) to the minimization of the function [44]

$$u \to \psi(u) := \Phi(u) - \int_{\Omega} f.u dx - \int_{\Gamma_1} g.\gamma_1(u)ds + \psi_C(u)$$

where the set C is defined by

$$C := \{u \in W^{1,p}(\Omega; \mathbb{R}^3) \mid \text{adj}(\nabla u) \in L^q(\Omega; \mathbb{R}^9), \det(\nabla u) \in L^r(\Omega; \mathbb{R}),$$
$$\det(\nabla u) > 0 \text{ a.e. in } \Omega, L(\nabla u) \leq 0 \text{ a.e. in } \Omega,$$
$$\gamma_2(u) \in Q^c \text{ on } \Gamma_2\}.$$

Here γ_i is the map wich associates $u \in W^{1,p}(\Omega; \mathbb{R}^3)$ with the restriction of the trace operator $\gamma : W^{1,p}(\Omega; \mathbb{R}^3) \to W^{1-\frac{1}{p},p}(\Gamma; \mathbb{R}^3)$ to Γ_i.

We assume that

$$C \cap dom(\Psi) \neq \emptyset. \tag{47}$$

Then by following a classical argumentation for the study of such problem [22], [44], [88] we can prove that ψ is weakly lower semicontinuous.

Proposition 4.5.1. ψ is weakly lower semicontinuous.

Proof: We prove that epi(ψ) is weakly closed. Let $\{u_n; n \in \mathbb{N}\}$ and $\{t_n; n \in \mathbb{N}\}$ be sequences sastisfying $u_n \to u, t_n \to t$ and $\psi(u_n) \leq t_n$. If $t = +\infty$ then the result is clear. If $t \in \mathbb{R}$ then

$$\limsup \psi(u_n) < +\infty. \tag{48}$$

Then assumption (42) implies that

$$\theta(u_n) := (u_n, \text{adj}(\nabla u_n), \det(\nabla u_n))$$

is bounded in $W^{1,p}(\Omega; \mathbb{R}^3) \times L^q(\Omega; \mathbb{R}^9) \times L^r(\Omega; \mathbb{R})$.

Therefore, by a result of J.M. Ball ([22]; Theorem 6.2), there exists a subsequence (again denoted by u_n) such that

$$\theta(u_n) \to \theta(u) \text{ in } W^{1,p}(\Omega; \mathbb{R}^3) \times L^q(\Omega; \mathbb{R}^9) \times L^r(\Omega; \mathbb{R}).$$

80

Thus $\gamma_2(u_n) \to \gamma_2(u)$ in $L^p(\Gamma; I\!R^3)$ and by considering a subsequence, we can assume that $\gamma_2(u_n)(x) \to \gamma_2(u)(x)$ a.e. on Γ_2. Then,

$$\gamma_2(u(x)) \in Q^c \text{ a.e. in } \Gamma_2,$$

since Q^c is closed.

By a result of I. Ekeland and R. Temam ([56]; Chapter VIII, Theorem 2.1), the functional $(F, H, \delta) \to \int_\Omega \mathbf{W}\,(F, H, \delta)dx$ is weakly lower semicontinuous on $L^p(\Omega; I\!R^9) \times L^q(\Omega; I\!R^9) \times L^r(\Omega; I\!R)$ and thus

$$\Phi(u) - \int_\Omega f.u\,dx - \int_{\Gamma_1} g.\gamma_1(u)ds$$
$$\leq \liminf(\Phi(u_n) - \int_\Omega f.u_n dx - \int_{\Gamma_1} g.\gamma_1(u_n)ds)$$
$$\leq \liminf \psi(u_n) \leq t,$$

so that $u \in \mathrm{dom}(\Phi)$ and therefore $\det(\nabla u) < +\infty$.

By Mazur's Theorem [56], there exists a convex combination

$$z_k := \sum_{l=0}^k \alpha_{lk}\theta'(u_l),$$

with $\theta'(u) := (\nabla u,\; adj(\nabla u),\; det(\nabla u))$ and

$$\sum_{l=0}^k \alpha_{lk} = 1, \alpha_{kl} \geq 0, 0 \leq 1 \leq k,$$

and such that $z_k \to \theta'(u)$. Thus, by considering a subsequence, we can assume that $z_k(x) \to \theta'(u)(x)$ a.e. on Ω. We have

$$\mathbf{L}(z_k) \leq \sum_{l=0}^k \alpha_{lk}\mathbf{L}(\theta'(u_l))$$

where $L(\nabla u) = \mathbf{L}(\theta'(u))$ and $\mathbf{L}$ is convex (polyconvexity of L). Thus

$$L(\nabla u) \leq 0, \text{a.e. on } \Omega.$$

Therefore $u \in C$.

Thus

$$\psi(u) \;=\; \Phi(u) - \int_{\Omega} f.u dx - \int_{\Gamma_1} g.\gamma_1(u) ds$$

$$\leq\; t.$$

∎

Theorem 4.5.1. If

$$\int_{\Omega} f.e dx + \int_{\Gamma_1} g.\gamma_1(e) ds \;<\; 0, \forall e \in \{Q^c\}_{\infty}, e \neq 0 \tag{49}$$

then

$$\mathrm{argmin}\{\psi(x) \mid x \in W^{1,p}(\Omega; I\!\!R^3)\} \neq \emptyset.$$

Proof: This result is a consequence of Corollary 3.1.3. Indeed, we have

$$\psi_{\infty}(w) \;\geq\; \Phi_{\infty}(w) - \int_{\Omega} f.w dx - \int_{\Gamma_1} g.\gamma_1(w) ds$$

$$\geq\; - \int_{\Omega} f.w dx - \int_{\Gamma_1} g.\gamma_1(w) ds,$$

for all element w such that $\theta(w) \in W^{1,p}(\Omega; I\!\!R^3) \times L^q(\Omega; I\!\!R^9) \times L^r(\Omega; I\!\!R)$.

If $w \in R(\Delta(\varepsilon_n))$ then by Proposition 3.1.1, there exists $u_n \in C$ such that $\| u_n \| \to +\infty, w_n := u_n / \| u_n \| \rightharpoonup w$ and

$$\limsup \psi(u_n) / \| u_n \|^p \;\leq\; 0. \tag{50}$$

Inequality (50) together with estimation (42) imply that

$$| \nabla w_n |_{0,p} \to 0.$$

Moreover $w_n \to w$ in $L^p(\Omega; I\!\!R^3)$ and thus $w_n \to w \in I\!\!R^3 \backslash \{0\}$. It is also clear that $w \in \{Q^c\}_{\infty}$ (see also Lemma 4.10.1) and all the assumptions of Corollary 3.1.3 are satisfied with $N := \{Q^c\}_{\infty}$.

∎

Remark 4.5.1. The model studied in this Section is due to P.G. Ciarlet and J. Nečas [44]. We refer to the article of F. Tomarelli [131] for the study of an hyperelastic three-dimensional body simply supported by a membrane.

4.6 Convergence of the energies and the calculus of variations involving convex functionals on cone constraints

Let us first consider the problem of finding a minimizer for the functional

$$\psi(u) := \int_\Omega F(\nabla u)dx - \int_\Omega f.udx, \tag{51}$$

on the set C defined by

$$C := \{u \in H^1(\Omega) \mid \gamma_0(u) = g \text{ on } \Gamma\}.$$

Here, Ω is an open bounded and regular (i.e. its boundary Γ is Lipschitz continuous and Ω is located on one side of Γ) set of $\mathbb{R}^n$, $g \in H^{\frac{1}{2}}(\Gamma)$ and $\gamma_0 : H^1(\Omega) \to H^{\frac{1}{2}}(\Gamma)$ denotes the trace operator. By surjectivity of γ_0, there exists $z \in H^1(\Omega)$ such that $\gamma_0(z) = g$. We assume that $F : \mathbb{R}^n \to \mathbb{R}$ is a smooth function (at least twice differentiable) and we suppose also that $f \in L^2(\Omega)$.

That means that by solving problem (51) we are able to get weak solutions for the Euler-Lagrange equation

$$-\text{div}(\nabla F(\nabla u)) = f \text{ in } \Omega,$$

$$u = g \text{ on } \Gamma.$$

Some results are known for this problem [57]. We propose to recover and generalize these results by using the recession analysis. We begin by recalling a well known result [57] which requires a coercivity condition on F.

Theorem 4.6.1. We assume that

i) F is convex and

ii) $F(p) \geq \alpha \mid p \mid^2 - \beta \ (\alpha > 0, \beta \geq 0), \forall p \in \mathbb{R}^n.$

Then problem (51) has at least one solution.

Proof: Since F is convex, the functional ψ turns out to be convex and weakly lower semicontinuous ([57]; Theorem 2.1). Moreover, we have

$$
\begin{aligned}
\psi(u) \ &= \ \int_\Omega F(\nabla u) - f.u \ dx \\
&\geq \ \alpha \parallel \nabla u \parallel_{0,2}^2 - \beta \mu(\Omega) - \parallel f \parallel_{0,2} \parallel u \parallel_{0,2} \\
&\geq \ \tfrac{1}{2}\alpha \parallel \nabla(u-z) \parallel_{0,2}^2 - \alpha \parallel \nabla z \parallel_{0,2}^2 - \beta \mu(\Omega) - \parallel f \parallel_{0,2} \parallel u \parallel_{0,2}
\end{aligned}
$$

so that by using Poincaré inequality, we obtain

$$
\begin{aligned}
\psi(u) \ &\geq \ c_1 \parallel u - z \parallel^2 - c_2 \parallel u \parallel - c_3 \ (c_1 > 0, c_2 \geq 0, c_3 \geq 0) \\
&\geq \ c_1 \parallel u \parallel^2 - c_2 \parallel u \parallel - c_3 - c_1 \parallel z \parallel^2 \ .
\end{aligned}
$$

Thus $\psi(u) \to +\infty$ as $\parallel u \parallel \to +\infty$ and we conclude by Theorem 3.1.1 and Remark 3.1.2 i).

∎

Let us now consider the more general problem of finding a minimizer for the functional

$$
\psi(u) \ := \ \int_\Omega F(\nabla u) - f.u \ dx, \tag{52}
$$

on some nonempty and weakly closed cone C.

If in addition C is convex then by solving problem (52) we will obtain solutions of the variational inequality

$$
u \in C : \int_\Omega \nabla F(\nabla u).(\nabla v - \nabla u)dx \ \geq \ \int_\Omega f.(v-u)dx, \forall v \in C
$$

84

or equivalently of the complementarity system

$$u \in C;$$

$$\int_\Omega \nabla F(\nabla u).\nabla v \; dx \; \geq \; \int_\Omega f.v \; dx, \forall v \in C;$$

and

$$\int_\Omega \nabla F(\nabla u).\nabla u \; dx = \int_\Omega f.u \; dx.$$

We set

$$\Phi(u) \; := \; \int_\Omega F(\nabla u)dx,$$

and we note that

$$\Phi_\infty(u) \; = \; \int_\Omega F_\infty(\nabla u)dx.$$

Indeed, let $u_o \in H^1(\Omega)$ be given. Because of the convexity of F, the function

$$f(x,t) \; := \; t^{-1}[F(\nabla u_o(x) + t\nabla u(x))] - F(\nabla u_o)]$$

is non-decreasing with respect to t, for almost all $x \in \Omega$. Then, the Beppo-Levi Theorem implies that

$$\Phi_\infty(u) \; = \; \int_\Omega F_\infty(\nabla u)dx.$$

We assume that the following assumptions hold:

Growth condition.

$$| F(p) | \leq \; C(1+ | p |^2), (C \; > \; 0)\forall p \in I\!\!R^n, \tag{53}$$

Uniformly strict convexity.

$$\zeta^T D^2 F(p)\zeta \; \geq \; \gamma | \zeta |^2, (\gamma \; > \; 0) \; \forall p, \zeta \in I\!\!R^n. \tag{54}$$

It has been proved by L.C. Evans ([57]; Theorem 2.2) that if F satisfies the conditions (53) and (54) then the convergence of energies improves weak to strong convergence, i.e. if

$$u_n \rightharpoonup u \text{ in } H^1(\Omega)$$

and

$$\Phi(u_n) \to \Phi(u)$$

then

$$u_n \to u \text{ in } H^1(\Omega).$$

In the following result, we point out the connection between the convergence of energies and our weak- asymptotically compactness condition.

Proposition 4.6.1. Suppose that conditions (53) and (54) are satisfied. We assume that

i) $\exists R > 0 : u \in C, \| u \| > R \Rightarrow \psi(u) \geq 0$;

ii) $F(0) = 0$.

Then $R(\Delta(\varepsilon_n))$ is a-compact.

Proof: Let $\{u_n; n \in I\!N\}$ be a sequence such that $u_n \in C, w_n := u_n / \| u_n \| \rightharpoonup w$, $\| u_n \| \to +\infty$ and

$$\psi(u_n) + \varepsilon_n \| u_n \|^2 \leq \psi(v) + \varepsilon_n \| v \|^2, \forall v \in C.$$

It is clear that $w \in C$ since C is a weakly closed cone. Therefore, by Proposition 3.1.1 (5), we have

$$\limsup \psi(u_n) \leq \psi(w)$$

86

By assumption (52), F is convex and the functional ψ turns out to be convex and weakly lower semicontinuous ([57]; Theorem 2.1).

For n great enough, we can assume that

$$\| u_n \| \geq \max\{1, R\}$$

and

$$\psi(u_n) \geq 0.$$

Thus for n great enough,

$$
\begin{aligned}
\psi(w_n) &= \psi(u_n \| u_n \|^{-1}) \\
&\leq \psi(u_n) \| u_n \|^{-1} + \psi(0)(1 - \| u_n \|^{-1}) \\
&\leq \psi(u_n).
\end{aligned}
$$

The last inequality is due to the fact that $\psi(0) = 0$, $\psi(u_n) \geq 0$ and $\| u_n \| \geq \max\{1, R\}$. Thus

$$
\begin{aligned}
\limsup \psi(w_n) &\leq \limsup \psi(u_n) \\
&\leq \psi(w).
\end{aligned}
$$

Since ψ is weakly lower semicontinuous, we have also

$$\psi(w) \leq \liminf \psi(w_n).$$

Thus $w_n \rightharpoonup w$, $\psi(w_n) \to \psi(w)$ and in particular

$$\Phi(w_n) \to \Phi(w).$$

With assumptions (53) and (54) on F, we know that the convergence of the energies Φ improves weak to strong convergence, i.e. $w_n \to w$ and the a-compactness of $R(\Delta(\varepsilon_n))$ follows.

■

Theorem 4.6.2. Suppose that assumptions (53) and (54) are satisfied. If

ı) $\exists R > 0 : u \in C, \| u \| > R \Rightarrow \psi(u) \geq 0$;

ii) $F(0) = 0$;

iii) $K_o(\psi_\infty) \cap C = \{0\}$.

Then problem (52) has at least one solution.

Proof: By Proposition 4.6.1, $R(\Delta(\varepsilon_n))$ is a-compact, and thus by Proposition 3.1.1, $R(\Delta(\varepsilon_n)) \subset K_o(\psi_\infty) \cap C \backslash \{0\}$. Assumption iii) implies that $R(\Delta(\varepsilon_n))$ is empty and we may conclude by using Theorem 3.1.1.

∎

Corollary 4.6.1. Suppose that assumptions (53) and (54) are satisfied. Moreover, if

i) $\exists R > 0 : u \in C, \| u \| > R \Rightarrow \Phi(u) \geq 0$;

ii) $F(p) \geq c, \forall p \in I\!\!R^n (c \in I\!\!R)$.

Then problem (51) has at least one solution for each $f \in L^2(\Omega)$ satisfying the condition

$$\int_\Omega f(x)v(x)dx < 0, \forall v \in C \backslash \{0\}. \tag{55}$$

Proof: It is clear that

$$\psi_\infty(u) \geq 0, \forall u \in C.$$

Thus

$$K_o(\psi_\infty) \cap C \subset C_o := \{u \in C : \Phi_\infty(u) = \int_\Omega f(x)u(x)dx\}.$$

Assumption ii) implies that $\Phi_\infty(u) \geq 0$. Therefore condition (55) implies that $C_o = \{0\}$. We conclude by application of Theorem 4.6.2.

∎

88

4.7 Minimization Theorems for homogeneous functionals on smooth constraint manifolds

The aim of this Section is to discuss the connection between the general recession analysis described in Section 3.1 and the approach of Vy Khoi Le and K. Schmitt [135].

Let X be a real reflexive Banach space, let Y be a Banach space with norm $\| \cdot \|_Y$ and suppose that $H : X \to Y$ is a strongly continuous mapping. Let $\sigma \in Y$ be fixed and let C be the weakly closed manifold given by

$$C := \{x \in X \mid H(x) = \sigma\}.$$

Let $\psi : X \to I\!R \cup \{+\infty\}$ be the functional defined by

$$x \to \psi(x) := \Phi(x) + j(x)$$

where $\Phi : X \to I\!R$,

$$x \to \Phi(x) := \langle Ax, x \rangle,$$

with $A : X \to X'$ such that the functional $x \to \Phi(x)$ is weakly lower semicontinuous. Moreover we suppose that there exists $c > 0$ and $0 \leq \beta < 2$ such that

$$\Phi(x) \geq -c \parallel x \parallel^\beta, \forall x \in C.$$

The functional $j : X \to I\!R \cup \{+\infty\}$ is assumed to be convex, lower semicontinuous and such that $j(0) = 0$. We assume that

$$\mathrm{dom}\{j\} \cap C \neq \emptyset.$$

It is clear that with these assumptions, all the hypothesis (h) required in Section 3.1 are satisfied.

A) THE CASE OF HILBERT SPACES X (= X').

Definition 4.7.1. ([135]; L.K. Vy - K. Schmitt) We say that the pair (Φ, j) has property (P) on C whenever the following hold: If $\{u_n; n \in I\!N\} \subset C$ is any sequence in C satisfying

$$\| u_n \| \rightarrow +\infty,$$
$$w_n := u_n / \| u_n \| \rightharpoonup 0,$$
$$\limsup \Phi(u_n) / \| u_n \|^2 \leq 0,$$

then there exists $v_o \in C$ such that

$$\limsup \psi(u_n) > \psi(v_o).$$

Let us now recall some conditions of K.L. Vy and K. Schmitt [135] guaranteeing that property (P) holds. Let K be a closed convex subset of X such that

$$C \subset K$$

and

$$0 \in K.$$

We set

$$j^+(v) := \max\{j(v), 0\}$$

and

$$S_1 := \{v \in X \mid \| v \| = 1\}.$$

Proposition 4.7.1. ([135]; K.L. Vy - K. Schmitt) Suppose that

1) $\langle Au, u \rangle \geq 0, \forall u \in K$;

ii) there exists $c > 0, P_o, P_1 : X \to I\!\!R^+$ such that

$$\lambda^{-1}\langle A(\lambda v), v \rangle + P_o(v) + P_1(v) + j^+(v) \geq c,$$

for each v satisfying

$$\lambda v \in K, \forall v \in K \cap S_1, \lambda \geq 1;$$

iii) $P_o(K)$ is bounded;

iv) there exists $s > 0$ such that

$$P_o(\lambda u) \leq \lambda^s P_o(u), \forall \lambda \in [0,1], \forall u \in K;$$

v) $v_n \in K, v_n \rightharpoonup 0 \Rightarrow P_1(v_n) \to 0$.

Then the pair (Φ, j) satisfies property (P) on C.

The following Proposition sets up the connection between the property (P) of the pair (Φ, j) and the property (Q) of $\psi = \Phi + j$.

Proposition 4.7.2. If the pair (Φ, j) satisfies property (P) on C then the functional ψ satisfies property (Q) on C.

Proof: Let $\{u_n; n \in I\!\!N\} \subset C$ be any sequence in C satisfying

$$\| u_n \| \to +\infty,$$

$$w_n := u_n / \| u_n \| \rightharpoonup 0,$$

and

$$\limsup \psi(u_n) / \| u_n \|^2 \leq 0. \tag{56}$$

Inequality (56) implies that

$$\limsup\{j(u_n)/\parallel u_n \parallel^2 + \Phi(u_n)/\parallel u_n \parallel^2\} \leq 0.$$

Since j is convex and $j(0) = 0$, we have also

$$\liminf j(u_n)/\parallel u_n \parallel^2 \geq \liminf j(u_n/\parallel u_n \parallel^2)$$
$$\geq 0.$$

Thus

$$\limsup \Phi(u_n)/\parallel u_n \parallel^2 \leq \limsup\{j(u_n)/\parallel u_n \parallel^2 + \Phi(u_n)/\parallel u_n \parallel^2\}$$
$$- \liminf j(u_n)/\parallel u_n \parallel^2 .$$
$$\leq 0. \tag{57}$$

Therefore there exists $v_o \in C$ such that

$$\limsup \psi(u_n) > \psi(v_o).$$

That means that Ψ has property (Q) on C. ∎

Remarks 4.7.1. i) Let us recall that if $w \in R(\Delta(\varepsilon_n))$ then by Proposition 3.1.1 there exists a sequence $\{u_n; n \in I\!N\}$ such that

(1) $u_n \in C$;

(2) $\parallel u_n \parallel \rightarrow +\infty$;

(3) $w_n := u_n.\parallel u_n \parallel^{-1} \rightharpoonup w \in w - C_\infty \cap K_o(w - \psi_\infty)$;

(4) $\limsup \psi(u_n)/\parallel u_n \parallel^p \leq 0, \forall p > 0$;

(5) if $v_o \in C$ then $\limsup \psi(u_n) \leq \psi(v_o)$.

Moreover we have here

(6) $\lim H(u_n)/\parallel u_n \parallel^\alpha = 0, \forall \alpha > 0.$

ii) Property (4) with $p = 1$ implies that

$$\limsup \Phi(u_n)/ \| u_n \| + j_\infty(w) \leq 0.$$

We now recover the basic existence result of K.L. Vy and K. Schmitt ([135]; Theorem 2.5) in Hilbert space.

Theorem 4.7.1. Let the pair (Φ, j) satisfy property (P) on C and suppose that the following compatibility condition is satisfied: If $w \in X$ is such that there exists a sequence $\{u_n; n \in I\!N\}$ satisfying

(1) $\quad \| u_n \| \rightarrow +\infty,$

(2) $\quad w_n := u_n. \| u_n \|^{-1} \rightharpoonup w,$

(3) $\quad \limsup \Phi(u_n)/ \| u_n \| + j_\infty(w) \leq 0,$

(4) $\quad \lim H(u_n)/ \| u_n \|^\alpha = 0, \forall \alpha > 0,$

then we have $w \in D_1$. Then there exists $u \in C$ such that

$$\psi(u) = \min\{\psi(v) \mid v \in C\}.$$

Proof: By Corollary 3.1.5, Proposition 4.7.2 and Remarks 4.7.1.

∎

B) THE CASE OF REAL REFLEXIVE BANACH SPACES.

Definition 4.7.2. ([135]; K.L. Vy - K. Schmitt) We say that the pair (Φ, j) has property (P^+) on C whenever the following hold. There exists a constant $p > 1$ such that: If $\{u_n; n \in I\!N\} \subset C$ is any sequence in C satisfying

$$\| u_n \| \rightarrow +\infty,$$

$$w_n := u_n/ \| u_n \| \rightharpoonup w,$$

$$\| u_n \| \leq \| u_n - \lambda w \|, \forall n, \forall \lambda \geq 1,$$

93

and

$$\limsup \Phi(u_n)/ \parallel u_n \parallel^p \leq 0,$$

then there exists $v_o \in C$ such that

$$\limsup \psi(u_n) > \psi(v_o).$$

Let us denote by $\Phi^{-1}(0)$ the set

$$\Phi^{-1}(0) = \{u \in X \mid \langle Au, u \rangle = 0\}.$$

Proposition 4.7.3. ([135]; K.L. Vy - K. Schmitt) Suppose that

i) $\Phi(u) \geq 0, \forall u \in X$;

ii) Φ is positively homogeneous of degree $p > 1$;

iii) $\Phi^{-1}(0)$ is a finite dimensional subspace of X;

iv) Φ is $\Phi^{-1}(0)$-invariant, i.e.

$$\Phi(u+v) = \Phi(v), \forall v \in X, \forall u \in \Phi^{-1}(0).$$

Furthermore assume that $X = \Phi^{-1}(0) \oplus X_o$, where X_o is a closed subspace of X and

v) $\Phi(v) \geq c \parallel v \parallel^p, \forall v \in X_o \ (p > 0)$.

Then the pair (Φ, j) satisfies property (P^+) on C.

As in Proposition 4.7.2 we prove that if the pair (Φ, j) satisfies property (P^+) on C then the functional ψ satisfies property (Q^+) on C, and by using Corollary 3.1.6 together with Remarks 4.7.1, we recover the existence Theorem of K.L. Vy and K. Schmitt ([135]; Theorem 4.4) in reflexive Banach space.

Proposition 4.7.4. If the pair (Φ, j) satisfies property (P^+) on C then the functional ψ satisfies property (Q^+) on C.

Theorem 4.7.2. Let ψ satisfy property (P^+) on C and suppose the following compatibility condition satisfied: If $w \in X$ is such that there exists a sequence $\{u_n; n \in \mathbb{N}\}$ such that

$$(1) \qquad \| u_n \| \to +\infty,$$

$$(2) \qquad w_n := u_n. \| u_n \|^{-1} \to w,$$

$$(3) \qquad \limsup \Phi(u_n) / \| u_n \| + j_\infty(w) \leq 0,$$

$$(4) \qquad \lim H(u_n) / \| u_n \|^\alpha = 0, \forall \alpha > 0,$$

then we have $w \in D_1$. Then there exists $u \in C$ such that

$$\psi(u) = \min\{\psi(v) \mid v \in C\}.$$

4.8 On the solvability of nonlinear partial differential equations

Many kinds of partial differential equations could also be studied by using the recession approach. An impressive number of papers using various approaches is devoted to this subject ([20], [21], [23], [25], [33], [38], [57], [63], [91], [97], [98], [100], [109] and [135]).

The aim of this Section is to present two problems chosen so as to show how the results of Section 3.1 can be applied for the study of partial differential equations.

A problem with exponential nonlinearity.

Let Ω be a bounded open and regular (i.e. with a C^1-boundary Γ and Ω located on one side of Γ) subset of $\mathbb{R}^2$. We are concerned with the following system [135] :

$$\Delta u + k e^u = h \text{ on } \Omega; \tag{58}$$

$$\frac{\partial u}{\partial n} = 0 \text{ on } \Gamma. \tag{59}$$

In order to find a weak solutions u^* of the system (58)-(59), that is $u^* \in H^1(\Omega)$ satisfying

$$\int_\Omega \nabla u^* . \nabla v \, dx + \int_\Omega h.v \, dx = \int_\Omega ke^{u^*} v \, dx, \forall v \in H^1(\Omega),$$

we consider the problem of finding a minimizer for the functional

$$\psi(u) := \frac{1}{2} \int_\Omega |\nabla u(x)|^2 \, dx + \int_\Omega h(x)u(x)dx \tag{60}$$

on the smooth and weakly closed "manifold" [135]

$$C := \{u \in H^1(\Omega) \mid \int_\Omega k(x)e^{u(x)}dx = 0\}.$$

We assume that

$$k \in C(\overline{\Omega}); \tag{61}$$

$$k \text{ assumes both positive and negative values;} \tag{62}$$

and

$$h \in L^2(\Omega). \tag{63}$$

Note that assumptions (61) and (62) ensure that C is nonempty.

Proposition 4.8.1. ψ satisfies property (Q) on C.

Proof: We set

$$\Phi(u) := \frac{1}{2} \int_\Omega |\nabla u(x)|^2 \, dx$$

and

$$j(u) := \int_\Omega h(x)u(x)dx.$$

We claim that all assumptions of Proposition 4.7.1 are satisfied with $K := H^1(\Omega), P_o(u) := 0$ and $P_1(u) = \frac{1}{2} \int_\Omega u^2(x)dx$. It is clear that assumptions i),

96

iii), iv) and v) are satisfied. It remains to prove that assumption ii) holds. Indeed, if $u \in H^1(\Omega), \| u \| = 1$ and $\lambda \geq 1$ then

$$\tfrac{1}{2}\lambda^{-1} \int_\Omega \nabla(\lambda u).\nabla u \, dx + P_1(u) + j^+(u) \geq \tfrac{1}{2}\int_\Omega | \nabla u |^2 \, dx + \tfrac{1}{2}\int_\Omega u^2 dx$$
$$= \frac{1}{2}\| u \|^2$$
$$= \tfrac{1}{2}.$$

The proof is complete.

∎

Let us now denote by Λ the set

$$\Lambda := \{u \in H^1(\Omega) \mid \int_\Omega \nabla u.\nabla v \, dx = \int_\Omega h.v \, dx, \forall v \in H^1(\Omega)\}.$$

If $\int_\Omega h(x)dx = 0$ then this set is nonempty as a consequence of the Fredholm alternative. Let $z \in \Lambda$ be given.

Theorem 4.8.1. Suppose that conditions (61)-(63) are satisfied. Moreover, we assume that

$$\int_\Omega h(x)dx = 0. \tag{64}$$

Then there exists $u \in C$ such that

$$\psi(u) \leq \psi(v), \forall v \in C.$$

Moreover, if

$$\int_\Omega k(x)e^{z(x)}dx < 0,$$

then there exists $u^* \in C$ such that

$$\int_\Omega \nabla u^*.\nabla v \, dx + \int_\Omega h.v \, dx = \int_\Omega k e^{u^*} v \, dx, \forall v \in H^1(\Omega).$$

Proof: If $w \in R(\Delta(\varepsilon_n))$ then Proposition 3.1.1(4) with $p = 2$, implies that

$$\limsup \psi(u_n)/ \| u_n \|^2 \leq 0$$

and thus

$$w \in \Phi^{-1}(0) = I\!R.$$

If $u \in C$ then $u - w \in C$ since

$$\int_\Omega k e^{u-w} dx \;=\; e^{-w} \int_\Omega k e^u dx$$
$$=\; 0.$$

Moreover

$$\psi(u - w) \;=\; \tfrac{1}{2} \int_\Omega |\nabla(u - w)|^2\, dx + \int_\Omega hu\, dx - \int_\Omega hw\, dx$$
$$=\; \tfrac{1}{2} \int_\Omega |\nabla(u)|^2\, dx + \int_\Omega hu\, dx$$
$$=\; \psi(u).$$

Therefore

$$R(\Delta(\varepsilon_n)) \subset D_1.$$

We conclude to the existence of a minimizer by application of Corollary 3.1.5. The second part of this Theorem follows from a result of K.L. Vy and K. Schmitt ([135]; Corollary 6.2).

A Dirichlet problem at resonance.

Let Ω be a bounded open and regular (i.e. with a C^1-boundary Γ and Ω located on one side of Γ) subset of $I\!R^2$. We consider the following problem:

$$f \in -\Delta u - \lambda_1 u + \beta(u) \text{ on } \Omega; \tag{65}$$

$$u \;=\; 0 \text{ on } \Gamma. \tag{66}$$

Here $f \in L^2(\Omega)$, β is a maximal monotone graph on $I\!R \times I\!R$ such that $D(\beta) = I\!R$, λ_1 is the principal eigenvalue of the Laplacian with respect to homogeneous Dirichlet

98

boundary data. It is well-known that λ_1 is a simple eigenvalue which is isolated and can be characterized by the formula

$$\lambda_1 = \inf\{\int_\Omega |\nabla u|^2 \, dx \mid u \in H_o^1(\Omega), \int_\Omega |u|^2 \, dx = 1\}.$$

If $E(\lambda_1)$ denotes the eigenspace corresponding to the eigenvalue λ_1 then we can write

$$E(\lambda_1) = \{\alpha.e_1 \mid \alpha \in \mathbb{R}\},$$

where e_1 is a corresponding eigenvector to λ_1 and which can be assumed positive on Ω.

It is well known that there exists a convex and lower semicontinuous function $j : \mathbb{R} \to \mathbb{R}$ such that

$$\partial j = \beta.$$

We set

$$\psi(u) := \frac{1}{2}(\int_\Omega |\nabla u|^2 - \lambda_1 |u|^2 \, dx) + \int_\Omega j(u)dx - \int_\Omega f(x)u(x)dx.$$

The variational form of (65)-(66) reads

$$\min\{\psi(u) \mid u \in H^1(\Omega)\}.$$

The following Theorem is a direct consequence of Corollary 3.1.4.

Theorem 4.8.2. If

$$\int_\Omega j_\infty(e)dx > \int_\Omega f(x)e(x)dx, \forall e \in E(\lambda_1)\backslash\{0\}. \tag{67}$$

Then there exists $u \in H^1(\Omega)$ such that

$$\psi(u) \leq \psi(v), \forall v \in H^1(\Omega).$$

99

If u is a minimizer for the functional ψ then

$$\int_\Omega j_\infty(e)dx \geq \int_\Omega f(x)e(x)dx, \forall e \in E(\lambda_1). \tag{68}$$

Remarks 4.8.1.

  i) (67) is equivalent to

$$-\int_\Omega j_\infty(-e_1)dx < \int_\Omega f(x)e_1(x)dx < \int_\Omega j_\infty(e_1)dx. \tag{69}$$

  ii) A condition like (67) is usually called a Landesman-Lazer condition [25], [91], [100].

4.9 On a class of semicoercive variational inequalities

Let X be a real Hilbert space $(X = X')$ and let us assume the following assumptions (t):

(t_1) $a : X \times X \to I\!R$ is a bounded linear symmetric and semicoercive form.

  Let us denote by $A : X \to X'$ the operator defined by

$$\langle Au, v \rangle = a(u, v), \forall u, v \in X.$$

  We assume also that

(t_2) $\dim\{Ker(A)\} < +\infty.$

(t_3) $H : X \to I\!R$ is a convex and strongly continuous functional $(x_n \rightharpoonup x \Rightarrow H(x_n) \to H(x))$;

(t_4) there exists $u^* \in X$ such that $H(u^*) < 0$;

(t_5) there exist $\alpha > 1, \alpha \neq 2$ such that

$$H(\lambda u) = \lambda^\alpha H(u), \forall \lambda \geq 0, \forall u \in X.$$

We are looking for solutions of the following variational inequality:

Find $u \in X$ such that :

$$u \notin Ker(A)$$

and

$$a(u, v - u) + H(v) - H(u) \geq 0, \forall v \in X. \tag{70}$$

Theorem 4.9.1. Suppose that assumptions $(t_1) - (t_5)$ are satisfied. Moreover, we assume that

(t_6) $H(u) > 0, \forall u \in Ker(A), u \neq 0.$

Then problem (70) has at least one solution.

Proof: We set

$$\psi(u) := \tfrac{1}{2}a(u, u)$$

and

$$C := \{u \in X \mid H(u) = -1\}.$$

Assumptions (t_4) and (t_5) imply that C is nonempty.

We claim that $R(\Delta(\varepsilon_n))$ is empty. Indeed, if $w \in R(\Delta(\varepsilon_n))$ then by Remark 4.7.1 i), there exists a sequence $\{u_n; n \in I\!N\}$ such that $\| u_n \| \to +\infty, w_n := u_n / \| u_n \| \to w,$

$$\limsup \psi(u_n)/ \| u_n \|^2 \leq 0, \tag{71}$$

and

$$\lim H(u_n)/ \| u_n \|^\alpha = 0. \tag{72}$$

Inequality (71) and assumptions (t_1) and (t_2) imply that (see the proof of Corollary 3.1.4)

$$w_n \to w \in Ker(A)\backslash\{0\}. \tag{73}$$

101

On the other hand (72) implies that

$$H(w) = 0.$$

Thus assumption (t_6) implies that $w = 0$, a contradiction to (73).

By Theorem 3.1.1, there exists $u \in C$ such that

$$\psi(u) = \min\{\psi(x) \mid x \in C\}.$$

It is clear that $u \notin Ker(A)$. Indeed, if we suppose the contrary then assumption (t_6) implies that either $H(u) > 0$ or $u = 0$. But since $u \in C$, we have also $H(u) = -1$ and a contradiction.

By the generalized Lagrange multipliers rule [48], there exists $(\theta, \mu) \in \mathbb{R}^2 \backslash (0,0)$ such that

$$0 \in \theta Au + \mu \partial H(u).$$

It is clear that $\theta \neq 0$. Indeed, if we suppose the contrary then we get

$$H(v) \geq -1, \forall v \in X.$$

which is a contradiction since assumption (t_5) implies that $H(2u) = -2^\alpha$ and $-2^\alpha < -1$. Therefore, we may assume without loss of generality that $\theta = 1$. Thus

$$0 \in Au + \mu \partial H(u),$$

that is ($\mu \neq 0$ since $u \notin KerA$)

$$\mu^{-1}a(u, v - u) + H(v) - H(u) \geq 0, \forall v \in X. \tag{74}$$

We set $v := \sqrt[\alpha]{2}u$ in (74) and we obtain

$$(1 - \sqrt[\alpha]{2})\mu^{-1}a(u, u) \leq -1 \tag{75}$$

so that $\mu > 0$.

If we set $u^* := \mu^{1/(\alpha-2)}u$ then we see that u^* is a solution of (70).

■

Theorem 4.9.2. Suppose that assumptions $(t_1) - (t_5)$ are satisfied. Moreover, we assume that

(t_7) H is $Ker(A)$-invariant, i.e.

$$H(u - z) = H(u), \forall u \in X, \forall z \in Ker(A).$$

Then problem (70) has at least one solution.

Proof: We define $\psi : X \to I\!\!R$ and C as in Theorem 4.9.1. By a similar argumentation that the one used in Theorem 4.9.1 we prove that

$$R(\Delta(\varepsilon_n)) \text{ is a} - \text{compact}$$

and

$$R(\Delta(\varepsilon_n)) \subset Ker(A)\backslash\{0\}.$$

Therefore if $w \in R(\Delta(\varepsilon_n))$ and $u \in C$ then $u - w \in C$ since

$$H(u - w) = H(u).$$

Moreover

$$\psi(u - w) = a(u - w, u - w)$$
$$= a(u, u)$$

so that

$$R((\Delta(\varepsilon_n)) \subset D_1.$$

By Corollary 3.1.1, there exists $u \in C$ such that

$$\psi(u) = \min\{\psi(x) \mid x \in C\}.$$

103

It is also clear that $u \notin Ker(A)$. Indeed, if we suppose the contrary then assumption (t_7) implies that

$$
\begin{aligned}
H(0) &= H(u - u) \\
&= H(u).
\end{aligned}
$$

However we have a contradiction since $H(0) = 0$ while $H(u) = -1$ since $u \in C$.

We conclude as in Theorem 4.9.1.

■

Example 4.9.1. We are concerned with the unilateral and periodic problem: Find $u \in H^1(\Pi)$ (here $\Pi = IR/TZ$, $T > 0$) such that

$$
\int_\Pi \dot{u}(\dot{v} - \dot{u})dt + \int_\Pi V(t, v)dt - \int_\Pi V(t, u)dt \geq 0, \forall v \in H^1(\Pi), \qquad (76)
$$

where

$$
\forall u \in IR, V(., u) \text{ is measurable and there exists } h \in L^1(0, T)
$$

such that $\forall t \in [0, T], \forall u \in IR, |u| = 1$, we have

$$
|V(t, u)| \leq h(t); \qquad (77)
$$

$\forall t \in IR, V(t, .)$ is continuous; $\qquad (78)$

$\exists \nu \in IR$ such that $V(., \nu) < 0$ on a non zero measure subset; $\qquad (79)$

$\exists \alpha > 1, \alpha \neq 2$ such that $\qquad (80)$

$$
V(., \lambda u) = \lambda^\alpha V(., u), \forall \lambda \geq 0, \forall u \in IR;
$$

$\forall t \in [0, T], V(t, .)$ is convex. $\qquad (81)$

The following Corollary generalizes to variational inequalities a result of A.K. Ben Naoum, C. Troestler and M. Willem [24] stated for differential equations.

Corollary 4.9.1. Suppose that conditions (77) - (81) hold true. Moreover, we assume that

$$
\forall e \in IR\backslash\{0\}, \int_\Pi V(t, e)dt > 0. \qquad (82)
$$

104

Then (76) has a nonconstant solution.

Proof: We set

$$\langle Au, v \rangle = \int_{\Pi} \dot{u}\dot{v}dt$$

and

$$H(u) = \int_{\Pi} V(t, u)dt.$$

Here $Ker(A) = I\!R$ and it is clear that assumptions $(t_1), (t_2), (t_5)$ and (t_6) are satisfied. The function $u \to H(u)$ is convex and its strong continuity follows from conditions (77), (78) and the dominated convergence Theorem. Indeed, let $u_n \rightharpoonup u$ in $H^1(\Pi)$. Then by considering a subsequence, we can assume that

$$u_n(t) \to u(t) \text{ a.e. in } \Pi.$$

Thus

$$V(t, u_n(t)) \to V(t, u(t)) \text{ a.e in } \Pi$$

and since $\mid V(t, u_n(t)) \mid \le h(t)$, we may conclude by using the dominated convergence Theorem. It remains to prove assumption (t_4). This can be done by following the argumentation of A.K. Ben Naoum, C. Troestler and M. Willem ([24]; Proposition 3.1).

We set

$$U := \{t \in \Pi \mid V(t, \nu) < 0\}.$$

Let us denote by χ_U the charasteristic function of U, and (ψ_ε) a mollifiers sequence, i.e. $\psi_\varepsilon \in C^\infty(\Pi), \text{supp}\{\psi_{\varepsilon\mid(-\frac{1}{2}T,\frac{1}{2}T)}\} \subset (-\varepsilon, +\varepsilon)$ and $\int_\Pi \psi_\varepsilon(t)dt = 1$ with $\psi_\varepsilon \ge 0$ on $I\!R$. It is well known that

$$\psi_\varepsilon * \chi_U \to \chi_U \text{ in } L^1(\Pi) \text{ as } \varepsilon \to 0.$$

Therefore, there exists a subsequence (again denoted by ψ_ε) such that

$$\psi_\varepsilon * \chi_U(t) \to \chi_U(t) \text{ a.e. in } \Pi$$

and thus

$$V(t,\nu)(\psi_\varepsilon * \chi_U(t))^\alpha \to V(t,\nu)(\chi_U(t))^\alpha = V(t,\nu)\chi_U(t).$$

Moreover

$$|V(t,\nu)(\psi_\varepsilon * \chi_U(t))^\alpha| = |V(t,\nu)| \left(\int_U \psi_\varepsilon(t-s)ds \right)^\alpha \leq h(t).$$

Thus

$$\int_\Pi V(t,\nu)(\psi_\varepsilon * \chi_U(t))^\alpha dt \to \int_U V(t,\nu)dt < 0.$$

Therefore if we choose ε small enough then

$$H(\nu(\psi_\varepsilon * \chi_U)) < 0.$$

∎

As a consequence of Theorem 4.9.2, we get also the following result.

Corollary 4.9.2. Suppose that conditions (77) - (81) are satisfied. Moreover, we assume that

$$\forall e \in I\!\!R \backslash \{0\}, \forall u \in H^1(\Pi), \tag{83}$$
$$\int_\Pi V(t, u+e)dt = \int_\Pi V(t,u)dt$$

Then (76) has a nonconstant solution.

Remark 4.9.1. See A.K. Ben Naoum, C. Troestler and M. Willem [24] and L.K. Vy and K. Schmitt [135] for similar results concerning partial differential equations. See D. Goeleven, V.H. Nguyen and M. Willem [73] where semicoercive variational inequalities are examined by using the nonsmooth critical point theory of A. Szulkin [129].

4.10 Hemivariational inequalities involving potential operators

The theory of variational inequalities is a well-developed theory in mathematics. From the mechanical point of view, it is well known that this theory is closely connected with the superpotential of J.J. Moreau [102], i.e. with the subdifferential operator of convex analysis (see Section 4.1 and Section 4.9).

In the case of lack of monotonicity of the underlying stress-strain or reaction-displacement conditions then another type of inequality expression arises as weak formulation of the problem. This variational expression is called hemivariational inequality and has been introduced by P.D. Panagiotopoulos [110]-[119] in order to study nonmonotone semipermeability problems, multilayered plates, composite structures, etc.

Let X be a real reflexive Banach space and let C be a nonempty weakly closed subset of X. We recall that we denote by $T_C(u)$ the Clarke's tangent cone of C at $u \in C$.

Let $f \in X'$ be given and let $A : X \to X'$ be a potential operator, i.e. there exists a functional $\Phi \in C^1(X, \mathbb{R})$ whose Fréchet derivative $\Phi'(u)$ has the property that

$$\langle \Phi'(u), v \rangle = \langle A(u), v \rangle, \forall u \in C, v \in V.$$

Let $J : X \to \mathbb{R}$ be a locally Lipschitz functional. Then the generalized directional derivative of J at x in the direction d is denoted by $J^0(x; d)$ and is defined by

$$J^0(x; d) := \limsup_{t \downarrow 0, h \to 0} \frac{J(x + h + td) - J(x + h)}{t}.$$

The generalized gradient of J at x is denoted by $\partial J(x)$ and is defined as the convex subdifferential of the convex function $d \to J^0(x; d)$. We refer the reader to the book of F. Clarke [48] for more details concerning nonsmooth analysis.

The problems related to the approach of P.D. Panagiotopoulos require the mathematical study of the following problem [108]:

Find $u \in C$ such that

$$\langle Au - f, v \rangle + J^0(u; v) \geq 0, \forall v \in T_C(u). \tag{84}$$

This formulation is called hemivariational inequality and has been introduced by P.D. Panagiotopoulos in order to study various mechanical problems in which it is necessary to introduce nonmonotone boundary conditions. Hemivariational inequalities involving potential operators can also be seen as generalized critical point problems, that is

$$u \in C : Au - f \in -N_C(u) - \partial J(u),$$

where $N_C(x)$ denotes the Clarke's normal cone to C at x. If $J \equiv 0$ then our problem reduces to the one we have examined in Section 3.3.

For instance, in order to illustrate the hemivariational inequality approach, we consider a material point with mass m which is constrained to remain in a closed subset of $I\!\!R^3$, for instance defined by

$$C := \{x \in I\!\!R^3 : f_1(x) \leq 0, f_2(x) \leq 0, ..., f_N(x) \leq 0\}$$

where the f_i's are continuously differentiable. Suppose that for $x . \in C$ some of the inequalities are satisfied as equalities. Let us denote their set by I. Then if $\mathrm{grad} f_i(x)$ $(i \in I)$ are linearly independent, it follows that one has

$$N_C(x) = \{v \in I\!\!R^3 : v = \sum_{i=1}^{N} \lambda_i.\mathrm{grad} f_i(x), \lambda_i \geq 0, i \in I\}.$$

As soon as the mass m is in contact without friction with the boundary of C, the reaction force is normal to the boundary, i.e.

$$R := -\sum_{i=1}^{N} \lambda_i.\mathrm{grad} \, f_i(x), \lambda_i \geq 0, f_i(x) \leq 0, \lambda_i.f_i(x) = 0.$$

That is,

$$x \in C \text{ and } R \in -N_C(x).$$

Therefore, if f_o is an external force acting on m, it is necessary and sufficient for the equilibrium that

$$x \in C \text{ and } f_o = -R \in N_C(x),$$

or also

$$x \in C \text{ and } \langle -f_o, v \rangle \geq 0, \forall v \in T_C(x).$$

If

$$f_o := m.g + f_1 + f_2,$$

where $m.g$ is the gravity force, f_1 is a force which derives from a potentiel, i.e. there exists a differentiable function $F : \mathbb{R}^3 \to \mathbb{R}$ such that

$$f_1 = -\text{grad } F(x),$$

and f_2 is a force which derives from a superpotential of P.D. Panagiotopoulos [110], i.e. there exists a locally Lipschitz function $J : \mathbb{R}^3 \to \mathbb{R}$ such that

$$f_2 \in -\partial J(x).$$

Then we get $u \in C$ and

$$\langle \text{grad } F(x) - mg, v \rangle + J^0(u; v) \geq 0, \forall v \in T_C(x).$$

This simple example shows that Problem P is nothing else than a general expression of the classical principle of virtual work [112]. Moreover, since A is a potential operator, a solution for problem (84) can be obtained by minimizing the 'energy' functional $x \to \Phi(x) + J(x) - \langle f, x \rangle$ on the set of 'admissible deformations' C.

- A Nonlinear Eigenvalue Problem for a Laminated Plate Problem.

We consider a laminated plate consisting of two isotropic and homogeneous laminae and the binding material between them [121]. We suppose that each lamina $\Omega^{(\alpha)} \subset I\!R^2 (\alpha = 1, 2)$ has a constant thickness $h^{(\alpha)}$ and buckles because of a boundary loading in the plate, of the form

$$\sigma_{ij}^{(\alpha)}.n_j^{(\alpha)} = \lambda.g_i^{(\alpha)} \text{ on } \Gamma^{(\alpha)} \tag{85}$$

where $n^{(\alpha)} \in I\!R^2$ denotes the outward normal unit vector on $\Gamma^{(\alpha)}, g^{(\alpha)} = g^{(\alpha)}(x)$ is a self-equilibrating compressive load distribution on the plate boundary $\Gamma^{(\alpha)}$ and λ is a real number measuring the magnitude of the lateral loading. Each lamina is identified with a bounded open and connected subset of $I\!R^2$ and its boundary is assumed appropriately smooth (a one-dimensional manifold of class C^1 such that $\Omega^{(\alpha)}$ is located in one side of $\Gamma^{(\alpha)}$). Let also the interlaminar binding material occupy a subset Ω' such that $\Omega' \subset \Omega^{(1)} \cap \Omega^{(2)}, \overline{\Omega}' \cap \Gamma^{(1)} = \emptyset$ and $\overline{\Omega}' \cap \Gamma^{(2)} = \emptyset$. The system is referred to a fixed right-handed cartesian coordinate system $Ox_1x_2x_3$ and the middle plane of each lamina coincides with the Ox_1x_2 plane. By $\zeta^{(\alpha)}(x)$ we denote the vertical deflection of the point $x = (x_1, x_2, x_3)$ of the α-th lamina and by $u^{(\alpha)} = (u_1^{(\alpha)}, u_2^{(\alpha)})$ the horizontal displacement of the α-th lamina.

The theory of Von Kármán plates gives rise to the following system of differential equations:

$$k^{(\alpha)} \Delta^2 \zeta^{(\alpha)} - h^{(\alpha)} (\sigma_{ij}^{(\alpha)} \zeta_{,j}^{(\alpha)})_{,i} = f_3^{(\alpha)} \text{ in } \Omega^{(\alpha)} \tag{86}$$

$$\sigma_{ij,j}^{(\alpha)} = 0 \text{ in } \Omega^{(\alpha)} \tag{87}$$

$$\sigma_{ij}^{(\alpha)} = C_{ijkl}^{(\alpha)} (\varepsilon_{ij}^{(\alpha)} + \tfrac{1}{2} \zeta_{,k}^{(\alpha)} \zeta_{,l}^{(\alpha)}) \text{ in } \Omega^{(\alpha)}. \tag{88}$$

Moreover, we assume that each lamina is clamped on a part $S^{(\alpha)}(H_2(S^{(\alpha)}) > 0)$ of its boundary and simply supported elsewhere, i.e.

$$\zeta^{(\alpha)} = 0 \text{ on } \Gamma^{(\alpha)} \text{ and } M^{(\alpha)}(\zeta^{(\alpha)}) = 0 \text{ on } \Gamma^{(\alpha)} \backslash S^{(\alpha)}, \tag{89}$$

$$\frac{\partial \zeta^{(\alpha)}}{\partial n} = 0 \text{ on } S^{(\alpha)}. \tag{90}$$

Here, $i, j, k, l = 1, 2, \Delta^2$ is the biharmonic operator, $k^{(\alpha)}$ is the bending rigidity of the α-th plate, $f^{(\alpha)} = (0, 0, F^{(\alpha)})$ is the distributed vertical load acting on the α-th lamina, and $M^{(\alpha)}$ is the bending moment of the α-th laminae. The tensors $\sigma^{(\alpha)} = \{\sigma_{ij}^{(\alpha)}\}$ and $\varepsilon^{(\alpha)} = \{\varepsilon_{ij}^{(\alpha)}\}$ denote the stress and strain tensor respectively in the plane of the α-th lamina and $C^{(\alpha)} := \{C_{ijkl}^{(\alpha)}\}$ is the corresponding elasticity tensor, the components of wich are assumed to be elements of $L^\infty(\Omega^{(\alpha)})$ and to satisfy the usual symmetry and ellipticity properties. Moreover, we assume that $g_i^{(\alpha)} \in L^2(\Gamma^{(\alpha)})$ and $F^{(\alpha)} \in L^2(\Omega^{(\alpha)})$.

Let $X \subset H^2(\Omega^{(1)}) \times H^2(\Omega^{(2)})$ be the Hilbert space defined by

$$X := X^{(1)} \times X^{(2)},$$

where (as usually, the boundary conditions are assumed to be satisfied in the trace sense)

$$X^{(\alpha)} = \{\zeta \in H^2(\Omega^{(\alpha)}), \zeta^{(\alpha)} = 0 \text{ on } \Gamma^{(\alpha)}, \frac{\partial \zeta}{\partial n}^{(\alpha)} = 0 \text{ on } S^{(\alpha)}\}.$$

More precisely, we set

$$X^{(\alpha)} := \{\zeta \in H^2(\Omega^{(\alpha)}), \gamma_0^{(\alpha)}(\zeta) = 0, \text{ a.e. on } \Gamma^{(\alpha)}, \gamma_{1|S^{(\alpha)}}^{(\alpha)}(\zeta) = 0 \text{ a.e. on } S^{(\alpha)}\},$$

where

$$\gamma^{(\alpha)} : H^2(\Omega^{(\alpha)}) \to W^{\frac{3}{2},2}(\Gamma^{(\alpha)}) \times W^{\frac{1}{2},2}(\Gamma^{(\alpha)}); \zeta \to \gamma^{(\alpha)}(\zeta) = (\gamma_0^{(\alpha)}(\zeta), \gamma_1^{(\alpha)}(\zeta))$$

is the trace map.

From equation (86), multiplying by $z^{(\alpha)} \in X^{(\alpha)}$, integrating and applying the Green-Gauss Theorem, we get the expressions

$$a(\zeta^{(\alpha)}, z^{(\alpha)}) + \int_{\Omega^{(\alpha)}} h^{(\alpha)} \sigma_{ij}^{(\alpha)} \zeta_{,i}^{(\alpha)} z_{,j}^{(\alpha)} dx = \int_{\Omega^{(\alpha)}} F^{(\alpha)} z^{(\alpha)} dx, \qquad (91)$$

where

$$a(\zeta^{(\alpha)}, z^{(\alpha)}) = k^{(\alpha)} \int_{\Omega^{(\alpha)}} (1 - \nu^{(\alpha)}) \zeta_{,ij}^{(\alpha)} z_{,ij}^{(\alpha)} + \nu^{(\alpha)} \Delta \zeta^{(\alpha)} \Delta z^{(\alpha)} dx.$$

111

Here $\nu^{(\alpha)} < \frac{1}{2}$ is the Poisson ratio of the α-th lamina. By (85) and the Green-Gauss Theorem, we get also

$$\int_{\Omega^{(\alpha)}} \sigma_{ij}^{(\alpha)} u_{i,j}^{(\alpha)} dx = \lambda \int_{\Gamma^{(\alpha)}} g^{(\alpha)} u^{(\alpha)} ds. \tag{92}$$

It is known from Von Kármán theory [47], [121] that for a homogeneous and isotrop plate, it is possible by using (92) to eliminate the variables $u^{(\alpha)}$ and split the tensor $\sigma^{(\alpha)}$ as follows

$$\sigma^{(\alpha)}(\zeta^{(\alpha)}) = \lambda . \theta^{(\alpha)} + \tau^{(\alpha)}(\zeta^{(\alpha)}),$$

where $\theta^{(\alpha)}$ does not depend of $\zeta^{(\alpha)}$ and $\tau^{(\alpha)}$ is a quadratic function of $\zeta^{(\alpha)}$.

Thus (91) reduces to

$$a(\zeta^{(\alpha)}, z^{(\alpha)}) \;+\; \lambda \int_{\Omega^{(\alpha)}} h^{(\alpha)} \theta_{ij}^{(\alpha)} \zeta_{,i}^{(\alpha)} z_{,j}^{(\alpha)} dx$$
$$+ \int_{\Omega^{(\alpha)}} h^{(\alpha)} \tau_{ij}^{(\alpha)} \zeta_{,i}^{(\alpha)} z_{,j}^{(\alpha)} dx = \int_{\Omega^{(\alpha)}} F^{(\alpha)} z^{(\alpha)} dx. \tag{93}$$

We put $F^{(\alpha)} = G^{(\alpha)} + R^{(\alpha)}$, where $G^{(\alpha)} \in L^2(\Omega^{(\alpha)})$ is the transversal load applied on the α-th lamina and $(R^{(1)}, R^{(2)}) \in L^2(\Omega^{(1)}) \times L^2(\Omega^{(2)})$ is a known function of $(\zeta^{(1)}, \zeta^{(2)})$ introduced so as to formulate the stress in the interlaminar binding layer Ω'.

Let C be a nonempty weakly closed subset of X. We will assume that

$$\zeta \in C.$$

For instance, if for each $x \in \Omega'$ we consider a nonempty closed subset $Q(x)$ of $I\!\!R^2$ and we suppose on the vertical deflection vector $\zeta(x)$ a general constraint of the form

$$\zeta(x) \in Q(x), \text{ a.e. on } \Omega',$$

then the set of admissible deflections is given by

$$C := \{\zeta \in X : \zeta(x) \in Q(x), \text{ a.e. on } \Omega'\}.$$

112

Let $h : \mathbb{R}^2 \to \mathbb{R}$ be a continuous function. By means of this possible nonconvex yield function, we can define a general nonconvex set of admissible deflections by setting

$$C = \{\zeta \in X : h(\zeta^{(1)}(x), \zeta^{(2)}(x)) \leq 0, \text{ a.e. on } \Omega'\}.$$

In each case, we will assume an abstract normality condition, i.e.

$$-R = -(R^{(1)}, R^{(2)}) \in N_C(\zeta),$$

where $R := (R^{(1)}, R^{(2)}) \in X' \times X'$ is the reaction force, i.e.

$$\langle R, z \rangle = \sum_{\alpha=1}^{2} \int_{\Omega^{(\alpha)}} R^{(\alpha)} z^{(\alpha)} dx, z \in X.$$

Thus

$$\langle R, h \rangle \geq 0, \forall h \in T_C(\zeta), \tag{94}$$

We obtain from (92) and (94) the following problem.

Find $\lambda \geq 0$ and $\zeta \in C$ such that

$$\begin{aligned}
\sum_{\alpha=1}^{2} a(\zeta^{(\alpha)}, z^{(\alpha)}) &+ \lambda . \sum_{\alpha=1}^{2} \int_{\Omega^{(\alpha)}} h^{(\alpha)} \theta_{ij}^{(\alpha)} \zeta_{,i}^{(\alpha)} z_{,j}^{(\alpha)} dx \\
&+ \sum_{\alpha=1}^{2} \int_{\Omega^{(\alpha)}} h^{(\alpha)} \tau_{ij}^{(\alpha)} \zeta_{,i}^{(\alpha)} z_{,i}^{(\alpha)} dx \\
&\geq \sum_{\alpha=1}^{2} \int_{\Omega^{(\alpha)}} G^{(\alpha)} z^{(\alpha)} dx, \forall z \in T_C(\zeta). \tag{95}
\end{aligned}$$

Let $G \in X'$ be defined by

$$\langle G, z \rangle = \sum_{\alpha=1}^{2} \int_{\Omega^{(\alpha)}} G^{(\alpha)} z^{(\alpha)} dx$$

and let us define the operators $J : X \to X', L : X \to X'$ and $T : X \to X'$ by the following formulae

$$\langle J\zeta, z \rangle = \sum_{\alpha=1}^{2} a(\zeta^{(\alpha)}, z^{(\alpha)});$$

$$\langle L\zeta, z \rangle = -\sum_{\alpha=1}^{2} \int_{\Omega^{(\alpha)}} h^{(\alpha)} \theta_{ij}^{(\alpha)} \zeta_{,i}^{(\alpha)} z_{,j}^{(\alpha)} dx;$$

113

and

$$\langle T\zeta, z \rangle \;=\; \sum_{\alpha=1}^{2} \int_{\Omega^{(\alpha)}} h^{(\alpha)} \tau_{ij}^{(\alpha)} \zeta_{,i}^{(\alpha)} z_{,j}^{(\alpha)} dx.$$

The properties of these operators are well known [46], [112] and we recall only some of them which will be used here. The operator J is bounded, symmetric, linear and coercive, L is linear, symmetric and compact, and T is completely continuous, cubic and positive, i.e. $\langle T\zeta, \zeta \rangle > 0$, for all $\zeta \neq 0$. Moreover T is a potential operator, more precisely $T\zeta = F'(\zeta)$ where $F(\zeta) = \langle T\zeta, \zeta \rangle / 4$.

We set

$$\Phi_\lambda(\zeta) := F(\zeta) + \tfrac{1}{2}\langle J\zeta, \zeta \rangle - \tfrac{1}{2}\lambda \langle L\zeta, \zeta \rangle.$$

and

$$\psi_\lambda(\zeta) \;=\; \Phi_\lambda(\zeta) - \langle G, \zeta \rangle.$$

Therefore, if ζ_λ is a minimizer for ψ_λ on C then $\zeta_\lambda \in C$ and

$$\langle J\zeta_\lambda - \lambda L\zeta_\lambda + T\zeta_\lambda - G, v - \zeta_\lambda \rangle \;\geq\; 0, \forall v \in T_C(\zeta_\lambda),$$

which is nothing else that the abstract formulation of (95).

Theorem 4.10.1. We assume that C is a nonempty weakly closed subset of X. For each $\lambda \geq 0$, problem (95) has at least one solution $u_\lambda \in C$ such that

$$\psi_\lambda(u_\lambda) \;\leq\; \psi_\lambda(v), \forall v \in C.$$

Proof: We claim that $R(\Delta(\varepsilon_n))$ is empty. Indeed, suppose the contrary and let u_o be any element of C. If $w \in R(\Delta(\varepsilon_n))$ then there exists a sequence $\{u_n; n \in I\!N\}$ such that $u_n \in C, t_n := \| u_n \| \to +\infty, w_n \rightharpoonup w$ and

$$\psi_\lambda(u_n) \;\leq\; c_o$$

114

for some constant c_o, that is

$$\frac{1}{2}\langle t_n J w_n - \lambda t_n L w_n, t_n w_n \rangle + t_n^4 F(w_n) \le c_o + \langle G, t_n w_n \rangle. \tag{96}$$

Since T is positive, (96) implies that

$$\frac{1}{2}\langle t_n J w_n - \lambda t_n L w_n, t_n w_n \rangle \le c_o + \langle G, t_n w_n \rangle$$

and thus

$$\alpha . t_n^2 \le \lambda t_n^2 \langle L w_n, w_n \rangle + 2c_o + 2\langle G, t_n w_n \rangle. \tag{97}$$

Dividing (97) by t_n^2 and taking the limit as $n \to +\infty$, we get

$$\alpha \le \lambda \langle L w, w \rangle. \tag{98}$$

If $\lambda = 0$, this is already a contradiction. If $\lambda > 0$ then (98) means that $w \ne 0$. Dividing (96) by t_n^4 and taking the limit, we get $F(w) = \langle Tw, w \rangle \le 0$. Since T is positive, that means that $\langle Tw, w \rangle = 0$ and thus $w = 0$, which is a contradiction.

∎

If $G \equiv 0$ and if $0 \in C$ then 0 is a trivial solution for problem (95) and it is worthwhile to discuss the existence of nontrivial solutions and a possible bifurcation from the line of trivial solutions.

We set

$$1/\rho := \sup\{\frac{\langle Lu, u \rangle}{\langle Ju, u \rangle} \mid u \in C \backslash \{0\}\}$$

and we assume that $0 < \rho < +\infty$. The following Theorem means that if the set of admissible elements satisfies some geometric assumptions then ρ is a critical load for our mechanical problem and if $\lambda > \rho$ then we can conclude to the existence of a post-buckling configuration.

We say that C is pseudo-convex with respect to $u_o \in C$ if

$$\{u_o\} \in \{u\} + T_C(u), \forall u \in C.$$

Theorem 4.10.2. Assume that $G = 0$ and $0 \in C$. If

115

i) C is weakly closed;

ii) C is pseudo-convex with respect to 0;

iii) $\alpha.C \subset C, \forall \alpha > 0$.

Then

(1) for each $\lambda \in [0, \rho], u = 0$ is the unique solution;

(2) for each $\lambda \in (\rho, +\infty)$, there exists at least one nontrivial solution;

(3) ρ is a bifurcation point, i.e. there exist sequences $\{\lambda_n; n \in I\!N\}$ and $\{u_n; n \in I\!N\}$ such that $u_n \neq 0, (\lambda_n, u_n)$ solution for problem (95), $u_n \to 0$ and $\lambda_n \to \rho$.

Proof:

(1) Let $\lambda \in [0, \rho]$ be given and suppose that $u \in C \backslash \{0\}$ is a solution for problem (95). We have

$$\langle Ju - \lambda Lu + Tu, v \rangle \geq 0, \forall v \in T_C(u).$$

By assumption ii) we have

$$0 \in u + T_C(u),$$

so that

$$\langle Ju, u \rangle - \lambda \langle Lu, u \rangle \leq -\langle Tu, u \rangle < 0. \tag{99}$$

If $\lambda = 0$ then we get

$$\alpha \parallel u \parallel^2 < 0,$$

which is a contradiction since $u \neq 0$. If $\lambda > 0$ then (99) implies

$$1/\lambda < \frac{\langle Lu, u \rangle}{\langle Ju, u \rangle} \leq 1/\rho,$$

so that $\lambda > \rho$, a contradiction.

(2) Let $\lambda > \rho$ be given. Theorem 4.10.1 gives the existence of a solution $u_\lambda \in C$ which is obtained as the minimum of the functional Φ_λ over C, i.e.

$$\Phi_\lambda(u_\lambda) \leq \Phi_\lambda(v), \forall v \in C.$$

Clearly, if there exists $z \in C$ such that $\Phi_\lambda(z) < \Phi_\lambda(0) = 0$ then the minimum is reached on $C\backslash\{0\}$. Since $\lambda > \rho$, there exist $e \in C$ such that

$$\langle Je, e \rangle - \lambda \langle Le, e \rangle < 0.$$

Suppose that $\Phi_\lambda(z) \geq 0$, for all $z \in C$, i.e.

$$F(z) + \tfrac{1}{2}\langle Jz, z \rangle - \tfrac{1}{2}\lambda\langle Lz, z \rangle \geq 0, \forall z \in C. \tag{100}$$

Put $z^* := t.e$, where $t > 0$. By assumption iii we have $z^* \in C$ and thus by using (100), we obtain

$$t^4 F(e) + \tfrac{1}{2}t^2\langle Je, e \rangle - \tfrac{1}{2}\lambda t^2\langle Le, e \rangle \geq 0$$

which implies that

$$t^2 F(e) + \tfrac{1}{2}\langle Je, e \rangle - \tfrac{1}{2}\lambda\langle Le, e \rangle \geq 0. \tag{101}$$

Taking the limit as t tends to 0^+ in (101), we get

$$\langle Je, e \rangle - \lambda \langle Le, e \rangle \geq 0,$$

a contradiction.

(3) Let $\{\lambda_n; n \in I\!N\}$ and $\{u_n; n \in I\!N\}$ be two sequences such that $\lambda_n \to \rho$, $\lambda_n > \rho$, $u_n \in C\backslash\{0\}$ and

$$\langle Ju_n - \lambda_n Lu_n + Tu_n, v \rangle \geq 0, \forall v \in T_C(u_n).$$

By assumption ii) we have

$$\langle Ju_n - \lambda_n Lu_n + Tu_n, u_n \rangle \leq 0. \tag{102}$$

117

We claim that there exists $\varepsilon > 0$ such that

$$\langle Tu_n, u_n \rangle \geq \varepsilon. \parallel u_n \parallel^4.$$

If this is not true then by considering eventually a subsequence and by setting $v_n := u_n / \parallel u_n \parallel$, we may suppose that

$$\lim_{n \to \infty} \langle T(v_n), v_n \rangle = 0,$$

and $v_n \to v_o$. Thus, since T is completely continuous, we would obtain $\langle Tv_o, v_o \rangle = 0$ and thus $v_o = 0$. By (102) we have

$$\lambda_n \langle Lu_n, u_n \rangle \geq \langle Ju_n, u_n \rangle + \langle Tu_n, u_n \rangle,$$

and thus

$$\lambda_n \langle Lu_n, u_n \rangle \geq \alpha \parallel u_n \parallel^2.$$

Dividing the last inequality by $\lambda_n \parallel u_n \parallel^2$, we get

$$\langle Lv_n, v_n \rangle \geq \alpha/\lambda_n.$$

Taking the limit as $n \to +\infty$, we obtain

$$0 = \langle Lv_o, v_o \rangle \geq \alpha/\rho > 0,$$

a contradiction. Therefore, by using (102) again we obtain

$$\langle Ju_n, u_n \rangle (1 - \lambda_n \rho^{-1}) + \varepsilon. \parallel u_n \parallel^4 \leq 0,$$

and thus

$$\varepsilon \parallel u_n \parallel^2 \leq \parallel J \parallel .(\lambda_n \rho^{-1} - 1). \tag{103}$$

Considering eventually a subsequence, we may suppose that u_n converges weakly to u^* and taking the limit $\lambda_n \to \rho$ in (103), we get $\varepsilon \parallel u^* \parallel^2 \leq 0$ and thus $u^* = 0$.

■

Remark 4.10.1. For instance, let C be defined by

$$C := \{u \in X : u^{(1)}.u^{(2)} \geq 0, \text{ a.e. on } \Omega'\}.$$

Then assumptions i, ii and iii of Theorem 4.10.2 are satisfied.

Claim i: C is weakly closed. Indeed, if $u_n \rightharpoonup u$ in C, then $u_n \to u$ a.e. on Ω' and $u_n^{(1)}(x).u_n^{(2)}(x) \geq 0$, a.e. on Ω'. Therefore $u^{(1)}(x).u^2(x) \geq 0$ on Ω' and then $u \in C$.

Claim ii: C is pseudo-convex with respect to 0. Indeed, we have to prove that if $u \in C$ then $-u \in T_C(u)$. Let $\{u_n; n \in I\!N\}$ and $\{\lambda_n; n \in I\!N\}$ be two sequences such that $u_n \in C, u_n \to u$ and $\lambda_n \downarrow 0$. We have to prove the existence of a sequence $k_n \to -u$ such that $u_n + \lambda_n.k_n \in C$. Put $k_n := -u_n$, we have

$$(u_n^{(1)}(x) + \lambda_n.k_n^{(1)}(x)).(u_n^{(2)}(x) + \lambda_n.k_n^{(2)}(x)) =$$

$$(1 - \lambda_n)^2 u_n^{(1)}(x).u_n^{(2)}(x) \geq 0 \text{ a.e. on } \Omega'.$$

Claim iii: $\alpha.C \subset C, \forall \alpha > 0$. Indeed, let $u \in C$ be given. We have $u^{(1)}.u^{(2)} \geq 0$, a.e. on Ω' and thus $\alpha^2 u^{(1)}.u^{(2)} \geq 0$, a.e. on Ω'. Therefore $\alpha.u \in C$.

- A nonconvex unilateral contact problem in elasticity.

Let Ω be a body identified as a bounded open subset of $I\!R^3$ referred to a coordinate system $\{0, x_1, x_2, x_3\}$ and Γ the body's surface supposed to be regular (i.e. Γ is an hypersurface of class $C^m (m \geq 1)$ and Ω is located on one side of Γ). It is assumed that Ω is subjected to a density force F. Surface tractions t are applied to an open portion Σ of Γ. The body Ω is assumed to be fixed along an open subset Γ_U of Γ (possibly empty).

Let $\sigma = \{\sigma_{ij}\}$ be the stress tensor and let $n = \{n_i\}$ be the outward unit normal vector on Γ. We denote by $S = \{S_i\}$ the stress vector on Γ, i.e. $S_i = \sigma_{ij}.n_j$.

Let u denotes the displacement field of the body. We consider the case of infinitesimal deformations of the body and we suppose that the body's material is characterized by a Cauchy elastic law, i.e. $\sigma_{ij} = C_{ijkl}.\varepsilon_{ij}$ where $\varepsilon = \{\varepsilon_{ij}(u)\}$ is the

119

strain tensor and $C = \{C_{ijkl}(x)\}$ is the linear-elasticity tensor. The elasticity tensor $C \in L^{\infty}(\Omega; I\!R^{81})$ is supposed to satisfy the classical symmetry properties :

$$C_{ijkl}(x) = C_{jikl}(x) = C_{ijlk}(x),$$

and the ellipticity property

$$C_{ijkl}(x)\zeta_{ij}\zeta_{kl} \geq m.\zeta_{ij}.\zeta_{kl},$$

for all $x \in \Omega$ and for all 3×3 symmetric matrices ζ.

We suppose that the body Ω is constrained to lie inside a nonempty closed and convex box $B \subset I\!R^3$, a.e. on Ω. We assume that $\overline{\Omega} \subset B$. The displacement field u satisfies the following system of equations:

$$\text{Equilibrium equation :} \quad -\partial\sigma_{ij}/\partial x_j = f_i + R_i \text{ in } \Omega, \tag{104}$$

$$\text{Constitutive equation :} \quad \sigma_{ij} = C_{ijkl}.\varepsilon_{ij} \tag{105}$$

$$\text{Boundary condition :} \quad u = 0 \text{ on } \Gamma_U, \tag{106}$$

$$S_i = t_i \text{ on } \Sigma, \tag{107}$$

$$\text{Unilateral condition :} \quad x + u(x) \in B \text{ a.e. in } \Omega. \tag{108}$$

Here, the reaction force R is introduced in order to describe the action of the constraints on the body. In order to formulate a frictionless contact between the body and the box, we introduce the multivalued law

$$-R(x) \in N_{Q(x)}(u(x)),$$

where $Q(x) = -x + B$. Equivalently, this amounts to say that

$$\psi_{Q(x)}(v) - \psi_{Q(x)}(u(x)) + R(x)(v - u(x)) \geq 0, \forall v \in I\!R^3, \text{ a.e. in } \Omega \tag{109}$$

On Σ we assume a subdifferential nonmonotone displacement-traction law [104] :

$$-t \in \partial j(x, u(x)) \tag{110}$$

where $j : \Sigma \times \mathbb{R}^3 \to \mathbb{R}$ is a locally Lipschitz function. Here $\partial j(x, \zeta)$ denotes the generalized Clarke's gradient of $j(x, .)$ with respect to ζ, i.e.

$$\partial j(x, \zeta) \ = \ \{w \in \mathbb{R}^3 : j^0(x, \zeta; w) \ \geq \ \zeta^T w, \forall w \in \mathbb{R}^3\}$$

where

$$j^0(x, \zeta; w) \ := \ \limsup_{y \to \zeta, t \downarrow 0} \frac{j(x, y + tw) - j(x, y)}{t}.$$

We assume that j satisfies the following conditions (k) :

(k_1) For all $\zeta \in \mathbb{R}^3$ the function

$$x \to j(x, \zeta)$$

is measurable on Σ;

(k_2) $j(x, 0) \ = \ 0, \forall x \in \Sigma$;

(k_3) There exists $\beta \in L^2(\Sigma, \mathbb{R}^3)$ such that

$$\mid j(x, y) - j(x, y') \mid \ \leq \ \beta(x). \mid y - y' \mid, \forall x \in \Sigma, \forall y, y' \in \mathbb{R}^3;$$

(k_4) $j(x, y) \ \geq \ c, \forall x \in \Sigma, y \in \mathbb{R}^3 (c \in \mathbb{R}_-)$.

Assumptions (k) mean that the function $J : L^2(\Sigma; \mathbb{R}^3) \to \mathbb{R}$ given by

$$J(v) \ = \ \int_\Sigma j(x, v(x)) ds, \tag{111}$$

is a Lipschitz continuous function defined onto $L^2(\Sigma; \mathbb{R}^3)$. Moreover, by using Fatou's Lemma, it is easy to see that [48]:

$$\int_\Sigma j^0(x, u(x); v(x)) ds \ \geq \ J^0(u; v), \forall u, v \in L^2(\Sigma; \mathbb{R}^3). \tag{112}$$

Let $\gamma : H^1(\Omega; \mathbb{R}^3) \to H^{\frac{1}{2}}(\Gamma; \mathbb{R}^3)$ be the continuous, linear and surjective trace operator. We denote by γ_U the map which associates $v \in H^1(\Omega; \mathbb{R}^3)$ with the restriction of $\gamma(v) \in H^{\frac{1}{2}}(\Gamma; \mathbb{R}^3)$ to Γ_U,

$$\gamma_U : H^1(\Omega; \mathbb{R}^3) \to H^{\frac{1}{2}}(\Gamma_U; \mathbb{R}^3).$$

121

Complementary to γ_U, we consider the trace operator

$$\gamma_\Sigma^0 : H^1(\Omega; I\!\!R^3) \to H^{\frac{1}{2}}(\Sigma; I\!\!R^3), \Sigma := \operatorname{int}(\Gamma \backslash \Gamma_U),$$

defined in a manner analogous to γ_U.

Let X be the Hilbert space defined by

$$X := \{v \in H^1(\Omega; I\!\!R^3) : \gamma_U(v) = 0 \text{ in } H^{\frac{1}{2}}(\Gamma_U; I\!\!R^3)\}.$$

and let C be the subset defined by

$$C := \{u \in X : u(x) \in Q(x), \text{ a.e. on } \Omega\}.$$

Remark that $0 \in C$ since $\overline{\Omega} \subset B$.

The mapping

$$\gamma_\Sigma^0 : X \to H_{00}^{\frac{1}{2}}(\Sigma; I\!\!R^3)$$

is continuous, linear and surjective [84].

Lemma 4.10.1. Let X be a real Banach space which is compactly embedded in $L^p(\Omega; I\!\!R^n)(n \in I\!\!N, n \geq 1, 1 < p < +\infty)$ where Ω is a bounded open subset of $I\!\!R^N(N \in I\!\!N, N \geq 1)$. Assume that for every $x \in \Omega, D(x)$ is a nonempty subset of $I\!\!R^n$ and let K be the subset of X defined by

$$K := \{u \in X : u(x) \in D(x), \text{ a.e. on } \Omega\}.$$

We suppose that $K \neq \emptyset$. Then

  i) $K_\infty \subset \{u \in X : u(x) \in D_\infty(x), \text{ a.e. on } \Omega\}$;

 ii) if for every $x \in \Omega, D(x)$ is closed in $I\!\!R^n$ then K is weakly closed in X;

iii) if for every $x \in \Omega, D(x)$ is closed and convex in $I\!\!R^n$ then K is closed convex in X and $K_\infty = \{u \in X : u(x) \in D_\infty(x), \text{ a.e. on } \Omega\}$.

Proof:

i) Let $v \in K_\infty$. There exist $\{t_n; n \in \mathbb{N}\}$ and $\{v_n; n \in \mathbb{N}\}$ such that $v_n \to v, t_n \to \infty$ and $t_n.v_n \in K$, that is

$$t_n.v_n(x) \in D(x), \text{ a.e. on } \Omega. \tag{113}$$

Since $X \hookrightarrow L^p(\Omega, \mathbb{R}^n)$ continuously, there exists a subsequence (again denoted by v_n) such that $v_n(x) \to v(x)$, a.e. on Ω. This together with (113) imply that $v(x) \in D_\infty(x)$ a.e. on Ω.

ii) Let $\{v_n; n \in \mathbb{N}\} \subset K$ be a sequence such that $v_n \rightharpoonup v$. Then $v_n \to v$ in $L^p(\Omega, \mathbb{R}^n)$ and for a subsequence $v_n(x) \to v(x)$ a.e. on Ω. Therefore

$$v_n(x) \in D(x), \text{ a.e. on } \Omega,$$

and thus $v \in K$ since $D(x)$ is closed.

iii) It is clear that K is convex and then closed by ii). Therefore K_∞ can also be written as follows

$$\cap_{\mu>0}(K-z)/\mu,$$

where z is any element of K. Let $v \in X$ be such that $v(x) \in D_\infty$, a.e. on Ω. Then, if $e \in K$, we obtain a.e. on Ω,

$$e(x) + \lambda v(x) \in D(x), \forall \lambda > 0.$$

Taking a sequence $\lambda_n \to +\infty$, we find that $e + \lambda_n v \in K$ and thus $v \in K_\infty$.

$\blacksquare$

A consequence of this Lemma is that C is a closed convex set and

$$C_\infty = \{u \in X : u(x) \in B_\infty, \text{ a.e. on } \Omega\}.$$

123

We assume that $F \in L^2(\Omega; \mathbb{R}^3), R \in L^2(\Omega; \mathbb{R}^3)$ and $t \in L^2(\Sigma; \mathbb{R}^3)$. Then (109) and (110) imply that

$$\int_\Omega R_i(v_i - u_i)dx \geq 0, \forall v \in C,$$

and

$$\int_\Sigma j^0(x, \gamma_\Sigma^0(u(x)); \gamma_\Sigma^0(v(x)))ds \geq -\int_\Sigma t_i \gamma_\Sigma^0(v)_i ds, \forall v \in X.$$

The functional $J(\gamma_\Sigma^0(.))$ is strongly continuous on $H^1(\Omega, \mathbb{R}^3)$. Indeed, if $u_n \rightharpoonup u$ and $v_n \rightharpoonup v$ in $H^1(\Omega, \mathbb{R}^3)$, then $\gamma_\Sigma^0(u_n) \rightharpoonup \gamma_\Sigma^0(u)$ and $\gamma_\Sigma^0(v_n) \rightharpoonup \gamma_\Sigma^0(v)$ in $H^{\frac{1}{2}}(\Sigma; \mathbb{R}^3)$. Therefore $\gamma_\Sigma^0(u_n) \to \gamma_\Sigma^0(u)$ and $\gamma_\Sigma^0(v_n) \to \gamma_\Sigma^0(v)$ in $L^2(\Sigma; \mathbb{R}^3)$. This together with the continuity of $J(.)$ on $L^2(\Sigma; \mathbb{R}^3)$ imply the strong continuity of $J(\gamma_\Sigma^0(u))$ as a function of u.

The principle of virtual work leads to the variational equality:

$$\int_\Omega -\sigma_{ij}(u)_{,j} v_i dx = \int_\Omega (F + R)_i v_i dx, \forall v \in X. \tag{114}$$

By using the Green's formula (if $\sigma_{ij}(u)_{,j} \in L^2(\Omega; \mathbb{R}^3)$) [84], that is:

$$\int_\Omega -\sigma_{ij}(u)_{,j}(v_i)dx + \langle \pi_\Sigma^0(\sigma(u)), \gamma_\Sigma^0(v) \rangle_\Sigma = \int_\Omega \sigma_{ij}(u)v_{i,j}dx,$$

where $\langle .,. \rangle_\Sigma$ denotes the duality pairing on $H_{00}^{\frac{1}{2}}(\Sigma; \mathbb{R}^3)' \times H_{00}^{\frac{1}{2}}(\Sigma; \mathbb{R}^3)$ and π_Σ^0 is a uniquely determined linear continuous mapping such that

$$\pi_\Sigma^0(\sigma(u))_i = \sigma_{ij}(u)_{|\Sigma} n_j \text{ if } \sigma_{ij}(u) \in C^1(\overline{\Omega}; \mathbb{R}^9).$$

By using condition (107) in the trace sense, we see that (114) is equivalent to

$$\int_\Omega \sigma_{ij}(u)(v_i)_{,j}dx = \int_\Omega (F + R)_i v_i dx + \int_\Sigma t_i v_i ds, \forall v \in X. \tag{115}$$

Then, taking in count the law on t and R, we get the hemivariational inequality

$$u \in C : a(u, v - u) + \int_\Sigma j^0(x, \gamma_\Sigma^0(u(x)); \gamma_\Sigma^0(v(x)) - \gamma_\Sigma^0(u(x)))ds$$

$$\geq \langle f, v - u \rangle, \forall v \in C, \tag{116}$$

124

where $a(u, v)$ is the bilinear continuous symmetric form

$$a(u, v) = \int_\Omega C_{ijhk} \varepsilon_{ij}(u) \varepsilon_{hk}(v) dx,$$

and $\langle f, v \rangle$ is the linear continuous form

$$\langle f, v \rangle = \int_\Omega F_i v_i dx.$$

A consequence of inequality (112) is that each solution of the abstract problem

$$u \in C : a(u, v - u) + J^0(\gamma_\Sigma^0(u); \gamma_\Sigma^0(v - u)) \geq \langle f, v - u \rangle, \ \forall v \in C, \qquad (117)$$

is a solution for problem (116) too. Moreover, by the chain rule [48], we have

$$\partial(J[\gamma_\Sigma^0])(u) \subset \partial J(\gamma_\Sigma^0(u))$$

and thus a solution for problem (116) can be obtained by minimizing the energy functional

$$\psi(u) = \tfrac{1}{2} a(u, u) + J(\gamma_\Sigma^0(u)) - \langle f, u \rangle$$

on the set C.

Let $A : X \to X'$ be the bounded linear and symmetric operator defined by

$$\langle Au, v \rangle = a(u, v), \forall u, v \in X.$$

It is known [112] that if $H_2(\Gamma_U) > 0$ then A is coercive. However if $\Gamma_U = \emptyset$ then A is only semicoercive and

$$Ker A = \{v \in X : v(x) = a \wedge x + b, a, b \in I\!\!R^3\}.$$

Therefore, a direct consequence of Corollary 3.1.4 is the following result.

125

Theorem 4.10.3.

i) If $H_2(\Gamma_U) > 0$ then problem (117) has at least one solution for each $f \in X'$.

ii) If $\Gamma_U = \emptyset$ then problem (117) has at least one solution for each $f \in X'$ satisfying the inequality

$$\langle f, w \rangle < 0, \forall w \in \{v \in H^1(\Omega; I\!\!R^3) \mid v(x) = a \wedge x + b \in B_\infty,$$
$$\text{a.e. in } \Omega, a, b \in I\!\!R^3, \mid a \mid + \mid b \mid \neq 0\}.$$

Proof: Condition (9) of Corollary 3.1.4 is satisfied since assumption (k_4) implies that $J_\infty \geq 0$.

■

Note that various other kinds of functions j that the ones satisfying conditions (k_1)-(k_4) could be examined in a similar way.

- Noncoercive hemivariational inequalities arising in nonlinear elasticity.

Let Ω be a bounded open connected subset of $I\!\!R^N$ with a boundary Γ that is Lipschitz continuous. The set $\overline{\Omega}$ is the reference configuration occupied by a homogeneous hyperelastic body in the absence of any applied force. We denote by $u : \overline{\Omega} \to I\!\!R^N$ the field of admissible deformations and by T the first Piola-Kirchhoff stress tensor $(T(F) = \frac{\partial W}{\partial F}(F) \in M^N)$.

We suppose that a part $\Omega' \subset \Omega$ of the body is constrained to remain in some given domain $Q \subset I\!\!R^N$. If we assume a contact without friction (when the body touch the boundary of the domain) then there is a reaction force which is normal to the boundary and the corresponding unilateral conditions can be described by the following two relations:

$$u(x) \in Q, \text{ a.e. on } \Omega' \tag{118}$$

126

and

$$- R \in N_Q(u(x)), \text{ a.e. on } \Omega', \tag{119}$$

where R denotes the reaction force. For various problems (especially the three-dimensional ones) it could be more convenient to consider (118) and (119) on a part of the boundary of the body. The treatment of this case needs some additional computations from the point of view of the formulation but the resulting mathematical model is of the same nature as the one derived here. For a deeper study of three-dimensional hyperelastic polyconvex materials with general nonmonotone unilateral boundary conditions, we refer the interested reader to the paper of D. Goeleven and P.D. Panagiotopoulos [76]. Here we limit our study to a model of relatively simplified but illustrative character.

The corresponding mixed displacement-traction problem is given as in Section 4.5 by the following model (we set $R(x) = 0$ for $x \in \Omega \backslash \Omega'$)

$$- \text{div } \mathbf{T}(\nabla u) = f + R \text{ in } \Omega; \tag{120}$$

and

$$T(\nabla u)n = g \text{ on } \Gamma. \tag{121}$$

The body forces f are given while for the surface forces, we assume a superpotential law of P.D. Panagiotopoulos, i.e. we suppose the existence of a locally Lipschitz function $j : \Gamma \times I\!R^N \to I\!R$ such that

$$-g \in \partial j(x, u(x)), \text{ a.e. on } \Gamma.$$

Then the boundary condition (121) reduces to

$$- T(\nabla u(x))n(x) \in \partial j(x, u(x)), \text{ a.e. on } \Gamma. \tag{122}$$

The law given by (122) is a general boundary condition which can be used to describe friction effects. See D. Goeleven and P.D. Panagiotopoulos [76] and the books of P.D. Panagiotopoulos [112] and [119] for more details. A basic situation is given when the following form holds for j :

$$j(z) \; = \; \sum_{i=1}^{N} \int_0^{z_i} \beta_i(\tau)d\tau, z \in I\!\!R^N,$$

and where β_i $(i = 1, ..., N)$ are locally bounded functions assuming good properties in order to have (see Section 4.12)

$$\partial j(z) \; = \; \prod_{i=1}^{N}[\underline{b}_i(z_i), \overline{b}_i(z_i)]$$

where

$$\underline{b}_i(z_i) \; = \; \lim_{\mu \to 0+} \underline{b}_{i\mu}(z_i)$$

and

$$\overline{b}_i(z_i) \; = \; \lim_{\mu \to 0+} \overline{b}_{i\mu}(z_i),$$

with

$$\underline{b}_{i\mu}(z_i) \; = \; \text{essinf } \{\beta_i(z_i) : | \zeta - z_i | < \mu\},$$

and

$$\overline{b}_{i\mu}(z_i) \; = \; \text{esssup } \{\beta_i(z_i) : | \zeta - z_i | < \mu\}.$$

We suppose that the data of the problem are sufficiently smooth in order to justify the following computations. From equation (120), we get

$$-\int_\Omega \text{div}\mathbf{T}(\nabla u).\theta dx \; = \; \int_\Omega f.\theta dx + \int_{\Omega'} R.\theta dx, \forall \theta \in X, \qquad (123)$$

where X denotes the space of admissible deformations (this space will be precised later in this section). By Green's Formula, we have

$$-\int_\Omega \text{div}\mathbf{T}(\nabla u).\theta dx \; = \; -\int_\Gamma \mathbf{T}(\nabla u)n.\theta dx + \int_\Omega \mathbf{T}(\nabla u) : \nabla \theta dx. \qquad (124)$$

128

The boundary condition (122) implies that

$$\int_\Gamma j^0(x, u(x); \theta(x)) dx \geq -\int_\Gamma \mathbf{T}(\nabla u) n.\theta dx, \forall \theta \in X. \tag{125}$$

Combining (123), (124) and (125), we get

$$\int_\Omega \mathbf{T}(\nabla u) : \nabla \theta dx + \int_\Gamma j^0(x, u(x); \theta(x)) dx$$
$$\geq \int_\Omega f.\theta dx + \int_{\Omega'} R.\theta dx, \forall \theta \in X. \tag{126}$$

The unilateral conditions (118) and (119) imply that

$$\int_\Omega R.\theta dx \geq 0, \forall \theta \in T(Q, u), \tag{127}$$

where

$$T(Q, u) := \{u \in X \mid u(x) \in T_Q(u(x)), \text{ a.e. on } \Omega'\}.$$

Therefore (126) leads us to the following inequality formulation of the principle of virtual work.

Find $u \in C := \{u \in X \mid u(x) \in Q, \text{ a.e. on } \Omega'\}$ such that

$$\int_\Omega \mathbf{T}(\nabla u) : \nabla \theta dx + \int_\Gamma j^0(x, u(x); \theta(x)) dx$$
$$\geq \int_\Omega f.\theta dx, \forall \theta \in T(Q, u). \tag{128}$$

If the following regularity property holds true

$$T(Q, u) \subset T_C(u), \tag{129}$$

then a solution for (128) can be obtained by solving the following hemivariational inequality: Find $u \in C$ such that

$$\int_\Omega T(\nabla u) : \nabla \theta dx + \int_\Gamma j^0(x, u(x); \theta(x)) dx$$
$$\geq \int_\Omega f.\theta dx, \forall \theta \in T_C(u). \tag{130}$$

129

Problem (128) is not suitable for a mathematical study and this is the reason why (see also [108]) we will be concerned with the solutions of problem (129). Note that the regularity condition (129) holds true in L^p-spaces (see E. Giner [66]; chapter VII, Proposition 1.5). Solutions of problem (130) will be called weak solutions for problem (128). Solutions of problem (130) can be obtained by minimizing the total energy functional whenever some suitable properties are satisfied by the functions W and j [48] and this rejoins in some sense the present approach with the one of P.G. Ciarlet and J. Nečas exposed in Section 4.5. Here, the total energy functional is given by the formula

$$u \to \psi(u) := \int_\Omega W(\nabla u)dx + \int_\Gamma j(x, u(x))dx + \int_\Omega f.udx$$

on the set C.

More precisely, we consider the problem of minimizing the functional

$$u \to \psi(u) := \int_\Omega W(\nabla u)dx + \int_\Gamma j(x, \gamma(u(x)))ds - \int_\Omega f.udx$$

on the set C defined by

$$C := \{u \in W^{1,p}(\Omega; I\!R^N) \mid u(x) \in Q \text{ a.e. on } \Omega\}.$$

Here $1 < p < +\infty, f \in L^q(\Omega; I\!R^N)(q^{-1} + p^{-1} = 1)$ and $\gamma : W^{1,p}(\Omega; I\!R^N) \to W^{1-\frac{1}{p},p}(\Gamma; I\!R^N)$ is the trace operator. We suppose that Q is closed and such that $\overline{\Omega'} \subset Q$.

We define the functional $J : W^{1,p}(\Omega; I\!R^N) \to I\!R$ by setting

$$J(u) := \int_\Gamma j(x, \gamma(u(x)))ds,$$

and we assume conditions on j such that J is well defined and the following inequality holds true [48]:

$$\int_\Gamma j(x, \gamma(u(x)); \gamma(v(x)))ds \geq J^0(u; v), \forall u, v \in W^{1,p}(\Omega; I\!R^N). \tag{131}$$

Therefore, a good way to solve our unilateral problem is to minimize the functional ψ on the set C. We have the following existence Theorem.

130

Theorem 4.10.4. Suppose that

i) the function

$$u \in W^{1,p}(\Omega; I\!\!R^N) \to I\!\!R; u \to \int_\Omega W(\nabla u)dx$$

is well-defined and weakly lower semicontinuous;

ii)

$$\int_\Omega W(\nabla u)dx \geq \alpha \parallel \xi(\nabla u) \parallel_{0,p}^p, \forall u \in W^{1,p}(\Omega; I\!\!R^N)$$

where $\alpha > 0$ and $\xi : I\!\!R^{N^2} \to I\!\!R^{N^2}$ is a positively homogeneous mapping such that the function

$$u \in W^{1,p}(\Omega; I\!\!R^N) \to I\!\!R; u \to \parallel \xi(\nabla u) \parallel_{0,p}^p$$

is weakly lower semicontinuous;

iii) if (compactness condition)

$$u_n \to u \text{ in } L^p(\Omega; I\!\!R^N)$$

and

$$\xi(\nabla u_n) \to 0 \text{ in } L^p(\Omega; I\!\!R^{N^2})$$

then

$$u_n \to u \text{ in } W^{1,p}(\Omega; I\!\!R^N);$$

iv) the function

$$u \in W^{1,p}(\Omega) \to I\!\!R, u \to J(u)$$

is weakly lower semicontinuous;

131

v)

$$J(u) \geq -\alpha_1 \| u \|^{\beta} -\alpha_2, \forall u \in W^{1,p}(\Omega; I\!\!R^N)$$

$$(\alpha_1 > 0, 0 < \beta < min\{2, p\}, \alpha_2 \in I\!\!R);$$

Then, if

$$\int_{\Omega} f.e dx < J_{\infty}(e), \forall e \in \Xi \cap C_{\infty} \backslash \{0\}$$

(here $\Xi := \{u \in W^{1,p}(\Omega; I\!\!R^N) \mid \xi(\nabla u) = 0\}$) then there exists

$$u \in \text{argmin}\{\int_{\Omega} W(\nabla u)dx + J(u) - \int_{\Omega} f.u dx \mid u \in C\}.$$

Proof: Let $w \in R(\Delta(\varepsilon_n))$. Then there exists a sequence u_n such that $w_n := u_n / \| u_n \| \rightharpoonup w$ in $W^{1,p}(\Omega; I\!\!R^N)$ and $\| u_n \| \to +\infty$. By Proposition 3.1.1, we get

$$\lim \inf \| \xi(\nabla w_n) \|_{0,p}^p \leq 0$$

and thus $\xi(\nabla w) = 0$.

Moreover, we have also

$$w_n \to w \text{ in } L^p(\Omega; I\!\!R^N)$$

and

$$\xi(\nabla w_n) \to 0 \text{ in } L^p(\Omega; I\!\!R^{N^2})$$

and thus

$$w_n \to w \in \Xi \cap C_{\infty} \backslash \{0\}.$$

We conclude by application of Corollary 3.1.3.

∎

Example 4.10.1. In linear elasticity, the model given for the stored energy is given by

$$W(\nabla F) = \tfrac{1}{2}\zeta(\nabla F) : E : \zeta(\nabla F)$$

with

$$\zeta(\nabla F) = \tfrac{1}{2}(\nabla F + \nabla F^T) - I$$

where $E = \{E_{ijkl}\} \in L^\infty(\Omega; I\!\!R^{81})$ is the Cauchy tensor whose coefficients are assumed to satisfy classical symmetry and ellipticity properties. If we set $z(x) = u(x) - x$ then the problem involves the energy of deformation given by

$$z \in W^{1,2}(\Omega; I\!\!R^3) \to I\!\!R, z \to \int_\Omega E_{ijkl}\epsilon_{ij}(z)\epsilon_{kl}(z)dx.$$

We have

$$\Xi = \{z \in W^{1,2}(\Omega; I\!\!R^3) \mid \epsilon(z) = 0\}$$

and assumptions (i-iii) of Theorem 4.10.4 are satisfied with ζ defined by $\zeta(M) = \tfrac{1}{2}(M + M^T)$.

Example 4.10.2. In nonlinear elasticity, a model for the stored energy, is for instance, the one given by the following formula

$$W(\nabla F) := \mid \nabla F \mid^p + \mid \text{adj}(\nabla F) \mid^{p/2},$$

where $p > 2$. In this case all assumptions of Theorem 4.10.4 are satisfied with $\xi(\nabla F) := \nabla F$. Here $\Xi = I\!\!R^3$ and the weak lower semicontinuity of the function

$$u \in W^{1,p}(\Omega; I\!\!R^3) \to I\!\!R; u \to \int_\Omega W(\nabla u)dx$$

is a consequence of a result of J.M. Ball ([22]; Corollary 6.2.2).

Remark 4.10.2. Suppose that $j : I\!\!R^3 \to I\!\!R$ is given by

$$j(z) = \sum_{i=1}^{3} \int_0^{z_i} \beta_i(\tau)d\tau$$

where $\beta_i \in L^\infty(I\!\!R)(i = 1, 2, 3)$. Then there exist two positive constants $C_1, C_2 > 0$ such that

$$-C_1 \mid z \mid \leq \mid j(z) \mid \leq C_2 \mid z \mid, \forall z \in I\!\!R^3$$

and

$$\mid j(z) - j(z') \mid \leq C_2 \mid z - z' \mid, \forall z, z' \in I\!\!R^3.$$

We end this Section by proving some necessary conditions concerning particular cases of our general problem. We set

$$K(u) := \int_\Omega W(\nabla u) dx$$

and analogously to (131) we assume conditions on W to have

$$\int_\Omega \mathbf{T}(\nabla u) : \nabla \theta dx \geq K^o(u; \theta), \forall u, v \in W^{1,p}(\Omega; I\!\!R^N).$$

We set

$$H := \{z \in W^{1,p}(\Omega, I\!\!R^N) : \int_\Omega \mathbf{T}(\nabla u) : \nabla z dx = 0, \forall u \in W^{1,p}(\Omega, I\!\!R^N)\}.$$

Note that $I\!\!R^N \subset H$.

Theorem 4.10.5. Suppose that

i) $x \to j(x, z)$ is measurable a.e. on Γ;

ii) $z \to j(x, z)$ is convex on $I\!\!R^N$;

iii) Q is a nonempty, closed and convex subset of $I\!\!R^N$.

If ψ has a minimum on C then

$$\int_\Gamma j_\infty(x, \gamma(e)) dx \geq \int_\Omega f.\gamma(e) dx, \forall e \in C_\infty \cap H.$$

134

Proof: Let u^* be a minimizer for ψ on C. The set C and the function $u \to J(u)$ are convex and thus

$$\int_\Omega \mathbf{T}(\nabla u^*) : (\nabla \theta - \nabla u^*)dx \quad + \quad \int_\Gamma j(x, \gamma(\theta(x)))dx - \int_\Gamma j(x, \gamma(u^*(x)))dx$$
$$\geq \quad \int_\Omega f.(\theta - u^*)dx, \forall \theta \in C.$$

Let $e \in C_\infty \cap H$. We set $\theta := u^* + e$ to obtain

$$\int_\Gamma j(x, \gamma(e(x) + u^*(x)))dx \quad - \quad \int_\Gamma j(x, \gamma(u^*(x)))dx \geq \int_\Omega f.edx.$$

By using Proposition 2.1.2 (2), we get

$$\int_\Gamma j_\infty(x, \gamma(e(x)))dx \geq \int_\Omega f.edx, \forall e \in C_\infty \cap H.$$

∎

Theorem 4.10.6. Suppose that

i)

$$j(x, z) = \sum_{i=1}^N \int_0^{z_i} \beta_i(x, \tau)d\tau, z \in I\!\!R^N, \text{ with } \beta_i(x, .) \in L^\infty(I\!\!R)$$

$(i = 1, ..., N)$ a.e. on Γ and $\beta_i(., z)(i = 1, ..., N)$ measurable, for all $z \in I\!\!R^N$;

ii) Q is a nonempty, closed and convex subset of $I\!\!R^N$.

Let $x \to \beta_{i,-}(x)$ and $x \to \beta_{i,+}(x)$ be measurable functions defined by $(i = 1, ..., N)$

$$\beta_{i,-}(x) \leq \beta(x, t) \leq \beta_{i,+}(x), \forall t \in I\!\!R, \text{ a.e. on } \Gamma.$$

Then if ψ admits at least one minimizer on C then

$$\text{(k)} \sum_{i=1}^N [\int_{\Gamma_+} \beta_{i,+}(x)\gamma(e(x))ds + \int_{\Gamma_-} \beta_{i,-}(x)\gamma(e(x))ds] \geq \int_\Omega f.edx,$$
$$\forall e \in C_\infty \cap H.$$

135

Here

$$\Gamma_+ := \{x \in \Gamma \mid \gamma(e(x)) > 0\}$$

and

$$\Gamma_- := \{x \in \Gamma \mid \gamma(e(x)) < 0\}.$$

Proof: We have

$$\int_\Omega \mathbf{T}(\nabla u^*) : \nabla(\theta - u^*)dx \quad + \quad \int_\Gamma j^0(x, \gamma(u^*(x)); \gamma(\theta(x) - u^*(x)))dx$$
$$\geq \quad \int_\Omega f.(\theta - u^*)dx, \forall \theta \in C.$$

Thus

$$\int_\Gamma j^0(x, \gamma(u^*(x)); \gamma(e(x)))dx \geq \int_\Omega f.edx, \forall e \in C_\infty \cap H.$$

We have

$$\int_\Gamma j^0(x, \gamma(u^*(x)); \gamma(e(x)))dx \quad = \quad \limsup_{t \downarrow 0, \varsigma \to u^*} \int_\Gamma \sum_{i=1}^N \int_{\gamma(\varsigma)}^{\gamma(\varsigma+te)} \beta_i(x, \tau)/t \, d\tau ds$$
$$\leq \quad \limsup_{t \downarrow 0, \varsigma \to u^*} \int_{\Gamma_+} \sum_{i=1}^N \int_{\gamma(\varsigma)}^{\gamma(\varsigma+te)} \beta_i(x, \tau)/t \, d\tau ds$$
$$+ \quad \limsup_{t \downarrow 0, \varsigma \to u^*} \int_{\Gamma_-} \sum_{i=1}^N \int_{\gamma(\varsigma+te)}^{\gamma(\varsigma)} -\beta_i(x, \tau)/t \, d\tau ds$$

Therefore

$$\int_\Gamma j^0(x, \gamma(u^*(x)); \gamma(e(x)))dx$$
$$\leq \sum_{i=1}^N [\int_{\Gamma_+} \beta_{i,+}(x)\gamma(e(x))ds + \int_{\Gamma_-} \beta_{i,-}(x)\gamma(e(x))ds].$$

∎

Remark 4.10.3.

i) Condition (k) can also be written as follows:

$$\sum_{i=1}^{N}[\int_{\Gamma}\beta_{i,+}(x)[\gamma(e(x))]_{+}ds - \int_{\Gamma}\beta_{i,-}(x)[\gamma(e(x))]_{-}ds] \geq \int_{\Omega}f.edx,$$

$$\forall e \in C_{\infty} \cap H.$$

Where for a function $x \to q(x)$, we denote its positive part by $[q(x)]_{+} := \sup\{q(x),0\}$ and its negative part by $[q(x)]_{-} = \sup\{-q(x),0\}$.

ii) Suppose that $C_{\infty} \cap H = I\!\!R$ and let us assume that

$$\beta_{i,-} \leq \beta(x,t) \leq \beta_{i,+}, \forall t \in I\!\!R, \text{ a.e. on } \Gamma.$$

Then condition (k) is equivalent to

$$\sum_{i=1}^{N}\beta_{i,+}H_{N-1}(\Gamma) \geq \int_{\Omega}f(x)ds \geq \sum_{i=1}^{N}\beta_{i,-}H_{N-1}(\Gamma).$$

- A robot hand grasping problem -

A robot hand grasping model [5], [128] is described in this Section. A more general theory involving problems which cannot be set in a variational form can be found in [128]. Let us consider a rigid object which is grasped by a robotic hand with n elastic fingers. The assumption of frictionless point contact between the fingertips and the rigid object allows us to consider only forces or reactions which the fingertips exert on the object with carrier line the normal to the boundary of the object through the point of contact. Let these forces be denoted by $r_i(i = 1, ..., n)$ and be assembled in a vector r.

The external forces or torques applied on the reference point of the rigid object are denoted by a vector $p = (p_1, p_2, p_3, m_1, m_2, m_3)^T$. A rectangular coordinate system $Ox_1x_2x_3$ is used. With respect to it, rigid body displacements and rotations are assembled in order to form the vector $u^0 := (u_1^0, u_2^0, u_3^0, \theta_1^0, \theta_2^0, \theta_3^0)^T$. Elements u_i^0 denote rigid body displacements while θ_i^0 denote rigid body rotations. The equilibrium equations for the rigid object subjected to external actions p and to finger-object reactions r are written in the form:

$$Gr = p, \qquad (132)$$

where G is the equilibrium matrix.

The boundary conditions in the direction normal to the boundary of the object are written in the form

$$y := u_N + d - u_N^0 = 0, \qquad (133)$$

where u_N is the vector of normal displacements of the fingertips, u_N^0 is the normal displacement of the object boundary at points adjacent to the fingertips and d is the possible nonzero initial gap between the fingertips and the object.

By using the principle of complementary virtual work, we can write

$$- u_N^0 = G^T u^0. \qquad (134)$$

Let contact forces and effective gaps be decomposed in a unilateral contact part and an adhesive part, i.e.

$$r = r_u + r_a$$

and

$$y = y_u + y_a.$$

Suppose that a frictionless unilateral contact gripper with no adhesion effects is considered (i.e. $r_a = 0, y_a = 0$). Unilateral contact effects between fingertips and object couple u_N and r_u are described as follows:

$$\text{if } u_N + d = u_N^0 \text{ then } r_u \geq 0$$

and

$$\text{if } u_N + d > u_N^0 \text{ then } r_u = 0.$$

138

In a more compact form, that is also

$$y = y_u \in -N_{I\!R_+^n}(r_u).$$ (135)

Let us now consider a linearly elastic finger with a symmetric flexibility matrix denoted by F, i.e.

$$u_N = Fr.$$ (136)

Using (134), (136) and the expression of y, we get

$$Fr + G^T u^0 + d = y_u.$$ (137)

We set

$$z = z_u := (r_u, u^0)^T,$$

$$M := \begin{pmatrix} F & G^T \\ \\ G & 0 \end{pmatrix}$$

and

$$b := (d, -p)^T.$$

Using (137), (132) and (135), we get the following variational inequality defined on $I\!R^{n+6}$ (three dimensional model):

$$z \in I\!R_+^n \times I\!R^6 : (Mz + b)^T(v - z) \geq 0, \forall v \in I\!R_+^n \times I\!R^6.$$ (138)

We are now in position to take into account locking effects with a general yield function that can be described by a nonempty closed subset D of $I\!R^n$. Here $r_a = 0$ and a nonconvex locking criterion is described by

$$y_a \in -N_D(r).$$ (139)

139

By adding 6 copies of the interval $(-\infty, +\infty)$ in the product space, i.e. we set

$$C := I\!R \times I\!R \times .... \times I\!R \times D.$$

Then relation (139) can be written in the space $I\!R^{n+6}$, i.e.

$$w_a \in -N_C(z). \tag{140}$$

Using (140) together with (138), we obtain the problem: Find $z \in C \cap I\!R_+^n \times I\!R^6$ such that

$$(Mz + b)^T(v - z) \geq 0, \forall (v - z) \in T_C(z) \cap I\!R_+^n \times I\!R^6.$$

Since, $I\!R_+^n \times I\!R^6 \cap T_C(z) \subset T_{(I\!R_+^{n+} \times I\!R^6) \cap C}(z)$, we can solve this problem by means of the hemivariational inequality

$$z \in E : (Mz + b)^T v \geq 0, \forall v \in T_E(z), \tag{141}$$

with

$$E := C \cap (I\!R_+^n \times I\!R^6).$$

Note that the model (141) can be used to study various kinds of unilateral problems in robotics, structural mechanics, plasticity, etc. (see [90], [108] and [128]).

Theorem 4.10.7. Suppose that $E \neq \emptyset$. If

$$-b^T e < \tfrac{1}{2} J_M(e), \forall e \in \{x \in E_\infty \backslash \{0\} \mid (Mx)^T x \leq 0\},$$

then problem (141) has at least one solution.

Proof: By using Proposition 3.1.1 (4), with $\psi(x) = \tfrac{1}{2}(Mx)^T x + b^T x$ we see that

$$R(\Delta(\varepsilon_n)) \subset \{x \in E_\infty \backslash \{0\} \mid (Mx)^T x \leq 0\}.$$

We conclude by application of Corollary 3.1.3 and Remark 3.1.2 (v).

■

Remark 4.10.4. i) If E is a closed cone then the condition of Theorem 4.10.7 can be relaxed by

$$-b^T e < \tfrac{1}{2} J_M^+(e), \forall e \in \{x \in E_\infty \backslash \{0\} \mid (Mx)^T x \le 0\}$$

where

$$J_M^+(e) = \liminf_{\substack{t \to +\infty \\ v \to e \\ v \in E}} t(Mv)^T v.$$

Indeed, in this case if $w \in R(\Delta(\varepsilon_n))$ then there exists a sequence $\{u_n, n \in I\!\!N\}$ such that $u_n \in E, \| u_n \| \to +\infty$, $w_n := u_n / \| u_n \| \to w$ and $\liminf \psi(u_n) / \| u_n \| \le 0$. Since E is a cone, we have $w_n \in E$ and thus

$$\tfrac{1}{2} J_M^+(w) + b^T w \le 0.$$

We know that $w \in \{x \in E_\infty \backslash \{0\} \mid (Mx)^T x \le 0\}$ and thus necessarily $R(\Delta(\varepsilon_n)) = \emptyset$. We conclude by application of Theorem 3.1.1. For instance, set $M = \left(\begin{smallmatrix} 1 & 2 \\ 2 & 0 \end{smallmatrix} \right)$ and $E = I\!\!R_+^2$. Then $\{x \in E_\infty \mid \langle Mx, x \rangle \le 0\} \backslash \{0\} = \{(x_1, x_2) \in I\!\!R^2 : x_1 = 0, x_2 > 0\}$. On this set $J_M^+ \equiv 0$ and the condition $b_2 > 0$ is a sufficient condition for the existence of a solution of problem (141). ii) If M is positive definite then $\{x \in E_\infty \backslash \{0\} \mid (Mx)^T x \le 0\} = E_\infty \cap Ker M \backslash \{0\}$ and $b^T e > 0, \forall e \in E_\infty \cap Ker M \backslash \{0\}$ is a sufficient condition for the existence of a solution of problem (141).

4.11 A nonvariational problem with nonconvex constraints

Let us consider a material point with mass m which is constrained to remain in a nonempty closed subset C of $I\!\!R^3$. When m is in contact without friction with the boundary of C then there is a reaction force R which is normal to the boundary, i.e.

$$R \in -N_C(x).$$

Therefore, if $f_o(x)$ is an external force acting on m, it is necessary and sufficient for the equilibrium that

$$x \in C \text{ and } f_o = -R \in N_C(x),$$

141

or also (here $\langle x, y \rangle := \sum_{i=1}^{3} x_i y_i$)

$$x \in C \text{ and } \langle -f_o(x), v \rangle \geq 0, \forall v \in T_C(x).$$

We suppose that

$$f_o(x) := mg + f(x),$$

where mg is the gravity force and where $f(x) = -Ax$ where $A \in \mathbb{R}^{3 \times 3}$ is a positive semidefinite matrix. Therefore, we have

$$\langle Ax - mg, v \rangle \geq 0, \forall v \in T_C(x).$$

Proposition 4.11.1. Let C be a nonempty subset of $\mathbb{R}^3$. The set-valued function $x \to \partial d_C(x)$ is pseudomonotone.

Proof: By ([48]; Proposition 2.1.2) the set $\partial d_C(x)$ is nonempty convex and compact. By ([48]; Proposition 2.1.5), the function $x \to \partial d_C(x)$ is upper semicontinuous. It remains to prove that if $u_n \to u$ and if $z_n \in \partial d_C(u_n)$ then for each $v \in \mathbb{R}^3$ there exists $z(v) \in \partial d_C(u)$ such that $\liminf \langle z_n, u_n - v \rangle \geq \langle z(v), u - v \rangle$. Let $v \in \mathbb{R}^3$ be given. We have

$$
\begin{aligned}
\liminf \langle z_n, u_n - v \rangle &= -\limsup \langle z_n, v - u_n \rangle \\
&\geq -\limsup d_C^0(u_n, v - u_n)
\end{aligned}
$$

The map $d_C^0(x, y)$ is upper semicontinuous as a function of (x, y) ([48], Proposition 2.1.1), and thus

$$\liminf \langle z_n, u_n - v \rangle \geq -d_C^0(u, v - u).$$

For each $y \in \mathbb{R}^3$ there exists $z(y) \in \partial d_C(u)$ such that ([48]; Proposition 2.1.2)

$$d_C^0(u, y) = \langle z, y \rangle.$$

142

Therefore, there exists $z(v) \in \partial d_C(u)$ such that

$$\liminf \langle z_n, u_n - v \rangle \geq \langle z(v), u - v \rangle.$$

∎

Since $x \to Ax$ is continuous, it is clear that $x \to (A + \lambda \partial d_C)(x)$ is pseudomonotone for each $\lambda \geq 0$. Therefore, we are in position to use the results presented in Section 3.3.

For instance, if C is closed and star-shaped with respect to $B(0, \rho)(\rho > 0)$, then by using Corollary 3.3.3, it is easy to see that

$$\langle g, v \rangle < 0, \forall v \in C_\infty \cap Ker(A + A^T) \backslash \{0\}$$

is a sufficient condition for the existence of an equilibrium.

Further applications in the field of adhesive grasping problems in robotics can be found in D. Goeleven, G.E. Stavroulakis and P.D. Panagiotopoulos [77]. Other applications in unilateral mechanics may also be found in D. Goeleven [72].

4.12 On partial differential equations with discontinuous nonlinearities

Partial differential equations involving discontinuous nonlinearities have now been studied in various directions by a quite number of mathematicians (see K.-C. Chang [41], [42], S. Carl [40] and D. Motreanu and Z. Naniewicz [103] where many references may be found).

However, we would like to Remark that most problems of partial differential equations involving discontinuous nonlinearities lead to hemivariational inequalities (see Section 4.10). However, there are several cases for which (for instance, if the discontinuous nonlinearities hold in the interior and not on the boundary) the hemivariational inequality is the only possible formulation of a mechanical problem in the chosen functional framework. In fact, variational inequalities are 'variational' formulations of partial differential equations involving convex and discontinuous nonlinearities, with possible infinite branches, while hemivariational inequalities are 'variational' formulations of partial differential equations involving possibly nonconvex discontinuous branches.

Let X be a real Hilbert space such that the embedding

$$X \hookrightarrow L^p(\Omega)$$

is dense and compact. Here $p > 1$ and Ω denotes a bounded domain of $\mathbb{R}^n$. Let us assume the following assumptions.

(l_1) $a : X \times X \to \mathbb{R}$ is a bounded linear symmetric and semicoercive form. As usually, let us denote by $A : X \to X'$ the operator defined by

$$\langle Au, v \rangle = a(u, v), \forall u, v \in X.$$

We assume also that

(l_2) $\dim\{Ker(A)\} < +\infty.$

(l_3) $H : X \to \mathbb{R}$ is the function defined by

$$H(u) = \int_\Omega \int_0^{u(x)} \beta(x, t) dt dx,$$

where β is a measurable function defined on $\Omega \times \mathbb{R}$.

(l_4) We suppose that $\beta(x, .) \in L^\infty_{\text{loc}}(\mathbb{R})$ and

$$| \beta(x, t) | \leq c_1 + c_2 | t |^{p-1} .$$

Then for each $x \in \Omega$, the function

$$t \to j(x, t) = \int_0^t \beta(x, \zeta) d\zeta$$

is locally Lipschitz on $\mathbb{R}$ and the function $u \to H(u)$ is locally Lipschitz on $L^p(\Omega)$ [42].

144

(l_5) We assume that the limits

$$\beta(x, \zeta \pm 0)$$

exist for every $\zeta \in \mathbb{R}, x \in \Omega$.

For $x \in \Omega$ be given, we set

$$\underline{b}(x, \zeta) \;=\; \lim_{\mu \to 0^+} \underline{b}_\mu(x, \zeta)$$

and

$$\overline{b}(x, \zeta) \;=\; \lim_{\mu \to 0^+} \overline{b}_\mu(x, \zeta)$$

with

$$\underline{b}_\mu(x, \zeta) \;=\; \operatorname{essinf} \{\beta(x, z) : |\zeta - z| < \mu\},$$

and

$$\overline{b}_\mu(x, \zeta) \;=\; \operatorname{esssup} \{\beta(x, z) : |\zeta - z| < \mu\}.$$

Then [42]

$$\partial j(x, \zeta) \;=\; [\underline{b}(x, \zeta), \overline{b}(x, \zeta)].$$

Recall that $\partial j(x, \zeta)$ denotes the subdifferential with respect to the second variable. Note also that here $\partial j(x, \zeta) = [\min\{\beta(x, \zeta - 0), \beta(x, \zeta + 0)\}, \max\{\beta(x, \zeta - 0), \beta(x, \zeta + 0)\}]$. Moreover, if the functions $\underline{b}(x, \zeta)$ and $\overline{b}(x, \zeta)$ are Baire-measurable and if $w \in \partial H(u)$ then we have ([42]; Theorem 2.1)

$$w(x) \in [\underline{b}(x, u(x)), \overline{b}(x, u(x))], \text{ a.e. on } \Omega. \tag{142}$$

From Corollary 3.1.4 we get the following abstract result applicable to eigenvalue problems involving discontinuous nonlinearities. We set

$$R(A, H_\infty) := \{w \in X \mid w \in Ker\{A\}, H_\infty(w) \leq 0, w \neq 0\}.$$

Theorem 4.12.1. Suppose that assumptions $(l_1) - (l_5)$ are satisfied. Moreover, we assume that

145

(l_6) there exists $e \in X$ such that $H(e) < 0$;

(l_7) for all $u \in X\backslash\{0\}$,

$$0 \notin [\underline{b}(x, u(x)), \overline{b}(x, u(x))] \text{ on a non zero measure subset of } \Omega;$$

(l_8) $R(A, H_\infty) \neq \emptyset$.

If $f \in L^q(\Omega) \, (q^{-1} + p^{-1} = 1)$ satisfies

$$\int_\Omega f.wdx < 0, \forall w \in R(A, H_\infty), \tag{143}$$

then there exists $\lambda \neq 0, u \in X\backslash\{0\}$ and $\chi \in X'$ such that

$$\langle Au - f, v \rangle = \lambda \int_\Omega \chi.v \, dx, \forall v \in X,$$

and

$$\chi(x) \in [\underline{b}(x, u(x)), \overline{b}(x, u(x))], \text{ a.e. on } \Omega.$$

Proof: We set

$$\psi(u) := \tfrac{1}{2}\langle Au, u \rangle - \int_\Omega f.udx$$

and

$$C := \{u \in X \mid H(u) = H(e)\}.$$

It is clear that this set in nonempty and weakly closed (it is a consequence of assumptions (l_3), (l_4), (l_6) and the compact embedding $X \hookrightarrow L^p(\Omega)$).

In order to apply Corollary 3.4.1, we have just to prove that

$$\langle f, w \rangle < 0, \forall w \in C_\infty \cap Ker(A)\backslash\{0\}. \tag{144}$$

If $w \in C_\infty$ then there exists a sequence $e_n \in C$ and a sequence $t_n \to +\infty$ such that $w_n := e_n/t_n \to w$. Therefore

$$H(t_n w_n)/t_n = H(e)/t_n$$

and thus

$$H_\infty(w) \leq 0.$$

Therefore assumption (143) implies condition (144) and by Corollary 3.1.4, there exist $u \in C$ such that

$$\psi(u) = \min\{\psi(x) \mid x \in C\}.$$

It is clear that assumption (l_6) implies that $u \neq 0$ since $H(u) = H(e) < 0$ and $H(0) = 0$.

Therefore, there exists $\alpha, \mu \in \mathbb{R}$ such that $\alpha.\mu \neq 0$ and

$$0 \in \alpha\psi'(u) + \mu\partial H(u). \tag{145}$$

We claim that $\mu \neq 0$. Indeed, if $\mu = 0$ then $\alpha \neq 0$ and $Au = f$ so that $\langle f, w \rangle = 0, \forall w \in Ker(A)$ and a contradiction to assumption (l_8) and condition (143). We have also $\alpha \neq 0$. Indeed, if $\alpha = 0$ then using (142), we get

$$0 \in [\underline{b}(x, u(x)), \overline{b}(x, u(x))] \text{ a.e. on } \Omega,$$

and a contradiction to assumption (l_7). The conclusion follows from (145) and (142).

■

147

Remarks 4.12.1

i) Suppose that the following limits exist:

$$\beta(x, -\infty) = \lim\{\bar{b}(x, t) \mid t \to -\infty\}$$

and

$$\beta(x, +\infty) = \lim\{\underline{b}(x, t) \mid t \to +\infty\}.$$

Then as a consequence of Fatou's Lemma and the generalized mean value Theorem [48], we get

$$H_\infty(w) \geq \int_{w>0} \beta(x, +\infty)w(x)dx + \int_{w<0} \beta(x, -\infty)w(x)dx.$$

Thus assumption (143) is satisfied if the following Landesman- Lazer condition is satisfied.

$$\int_{w>0} \beta(x, -\infty)w(x)dx \;+\; \int_{w<0} \beta(x, +\infty)w(x)dx \;<\; \int_\Omega f(x)w(x)dx$$
$$<\; \int_{w>0} \beta(x, +\infty)w(x)dx + \int_{w<0} \beta(x, -\infty)w(x)dx,$$
$$\forall w \in Ker(A)\backslash\{0\}.$$

ii) See K.-C. Chang [41], [42], D. Motreanu and Z. Naniewicz [103], D. Goeleven, D. Motreanu and P.D. Panagiotopoulos [78], Z. Naniewicz [107] for other results concerning this kind of problem.

4.13 An ill-posed linear system

We end this work with a very simple problem in order to recall that the solution obtained by our recession approach is given as the limit of a sequence $u_n \in \Delta(\varepsilon_n)$ and satisfies a viscosity selection principle.

Let A be a real matrix of order N. Given $b \in \mathbb{R}^N$, a solution of the linear equation

$$Ax = b \tag{146}$$

can be obtained by solving the quadratic minimization problem

$$\min\{\psi(x) \mid x \in I\!\!R^N\},$$

with

$$\psi(x) := \|Ax - b\|^2.$$

Using Proposition 3.2.1 ((4) with $p = 2$), it is easy to see that

$$R(\Delta(\varepsilon_n)) \subset Ker(A)\backslash\{0\}.$$

Therefore

$$R(\Delta(\varepsilon_n)) \subset D_1.$$

Using Corollary 3.2.1, we get the existence of a solution u which is obtained as the limit of a sequence u_n satisfying

$$(A^T A + \varepsilon_n I)u_n = A^T b. \tag{147}$$

Moreover, we have

$$\|u\| \le \|v\|, \forall v \in \operatorname{argmin}\{\psi(x) \mid x \in I\!\!R^N\}. \tag{148}$$

149

5 BIBLIOGRAPHY

References

[1] E. ACERBI and N. FUSCO. Semicontinuity Problems in the Calculus of Variations. *Arch. Rational Mech. Anal.*, 86, pp. 125-145, 1984.

[2] R. ADAMS. Sobolev Spaces. *Academic Press, New York*, 1975.

[3] S. ADLY, D. GOELEVEN and M. THERA. Recession Mappings and Noncoercive Variational Inequalities. *Nonlinear Analysis, T.M.A.*, Vol. 26, pp. 1573-1603, 1996.

[4] S. ADLY, D. GOELEVEN and M. THERA. Recession Methods in Monotone Variational Hemivariational Inequalities. *Topological Methods in Nonlinear Analysis*, Vol. 5, pp. 397-409, 1995.

[5] A.M. Al FAHED, G.E. STAVROULAKIS, P.D. PANAGIOTOPOULOS. Hard and Soft Fingered Robot Grippers. The Linear Complementary Approach. *Z.A.M.M.*, 71, pp. 257-265, 1991.

[6] D.D. ANG, K. SCHMITT and L.K. VY. Noncoercive Variational Inequalities: Some Applications. *Nonlinear Analysis, Theory, Methods & Applications*, Vol. 15, N°6, pp. 497-512, 1990.

[7] D.D. ANG, K. SCHMITT and L.K. VY. Variational Inequalities and the Contact of Elastic Plates. *In Differential Equations with Applications in Biology, Physics, and Engineering, Marcel Dekker, Inc. New York*, 1991.

[8] D.D. ANG, K. SCHMITT and L.K. VY. P-coercive Variational Inequalities and Unilateral Problems for von Kármán's Equations. *WSSIAA*, 1, pp. 15-29, 1992.

[9] G. ANZELLOTTI. A Class of Convex Non-Coercive Functionals and Masonry-Like Materials. *Ann. Inst. Henri Poincaré*, Vol. 3, N°4, p. 261-307, 1985.

[10] G. ANZELLOTTI, G. BUTTAZZO and G. DAL MASO. Dirichlet Problem for Demi-Coercive Functionals. *Nonlinear Analysis, Theory, Methods & Applications*, Vol. 10, N°6, pp. 603-613, 1986.

[11] H. ATTOUCH. Variational Convergence for Functions and Operators. *Pitman Publishing Inc*, 1984.

[12] H. ATTOUCH. Convergence of Viscosity Type Approximation Methods to Particular Solutions of Variational Problems. *In Proceedings of Variational Methods Nonlinear Analysis and Differential Equations, Genova Nervi, September 1993. E.C.I.G.-Genova*, pp. 13-43, 1994.

[13] H. ATTOUCH and M. THERA. A General Duality Principle for the Sum of two Operators. *To appear in Journal of Convex Analysis.*

[14] H. ATTOUCH, Z. CHBANI and A. MOUDAFI. Recession Operators and Solvability of Variational Problems. *Laboratoire d'Analyse Convexe, Université de Montpellier*, Preprint 1994/05.

[15] J.P. AUBIN. Applied Functional Analysis. *J. Wiley and Sons, New York*, 1979.

[16] J.P. AUBIN and H. FRANKOWSKA. Set-valued Analysis. *Birkhäuser Boston Inc., Cambridge*, 1990.

[17] J.P. AUBIN and F. CLARKE. Shadow Prices and Duality for a Class of Optimal Control Problems. *SIAM J. Control and Optimization*, 17, pp. 567-586, 1979.

[18] A. AUSLENDER. Noncoercive Optimization Problems. *To appear in Mathematics of Operations Researchs.*

[19] C. BAIOCCHI and A. CAPELO. Variational and Quasivariational Inequalities, Applications to Free-Boundary Problems. *John Wiley and Sons, New York,* 1984.

[20] C. BAIOCCHI, F. GASTALDI and F. TOMARELLI. Some Existence Results on Noncoercive Variational Inequalities. *Annali Scuola Normale Superiore - Pisa, Classe di Scienze,* Serie IV - Vol. XIII, N° 4, 1986.

[21] C. BAIOCCHI, G. BUTTAZZO, F. GASTALDI and F. TOMARELLI. General Existence Theorems for Unilateral Problems in Continuum Mechanics. *Arch. Rat. Mech. Anal.,* Vol. 100, N°2, pp. 149-180, 1988.

[22] J.M. BALL. Convexity Conditions and Existence Theorems in Nonlinear Elasticity. *Arch. Rational Mech. Anal.,* 63, pp. 337-406, 1977.

[23] P. BENILAN, M. CRANDALL and P. SACKS. Some Existence and Dependence Results for Semi-Linear Equations with Nonlinear Boundary Conditions. *Appl. Math. Optim.,* 17, pp. 203-224, 1988.

[24] A.K. BEN NAOUM, C. TROESTLER and M. WILLEM. Existence and Multiplicity Results for Homogeneous Second Order Differential Equations. *J. Diff. Equations,* 112, pp. 239-249, 1994.

[25] L. BOCCARDO, P. DRABEK and M. KUCERA. Landesman-Lazer Conditions for Strongly Nonlinear Boundary Value Problems. *Comment. Math. Univ. Carolinae,* 30, 3, pp. 411-427, 1989.

[26] G. BOUCHITTE and P. SUQUET. Equicoercivity of Variational Problems: the role of recession functions. *Collège de France, Pitman Research Notes in Maths,* 1994.

[27] M. BOUKROUCHE and G. BAYADA. The Characteristics Methods to Solve a Stationary Semi-Coercive Free Boundary Problem of Hydrodynamic Lubrication with Cavitation and Subject to an Integral Condition. *Journal of Mathematical Analysis and Applications*, 181, pp. 816-835, 1994.

[28] N. BOURBAKI. Eléments de Mathématiques - Espaces vectoriels Topologiques. *Hermann, Paris*, 1966.

[29] H. BREZIS. Equations et Inéquations Non Linéaires dans les Espaces Vectoriels en Dualité. *Ann. Inst. Fourier, Grenoble*, 18, 1, 115-175, 1968.

[30] H. BREZIS. Propriétés Régularisantes de Certains Semi-Groupes Non Linéaires. *Israël J. Math.*, Vol. 9, pp. 513-534, 1971.

[31] H. BREZIS. Problèmes Unilatéraux. *J. Math. Pures et Appl.*, 51, pp. 1-168, 1972.

[32] H. BREZIS. Opérateurs Maximaux Monotones et Semi-Groupes de Contractions dans les Espaces de Hilbert. *North-Holland*, 1973.

[33] H. BREZIS and L. NIRENBERG. Characterizations of the Ranges of Some Nonlinear Operators and Applications to Boundary Value Problems. *Ann. Scuola Normale Superiore Pisa, Classe di Scienze*, Serie IV, vol. V, N°2, pp. 225-326, 1978.

[34] H. BREZIS and A. HARAUX. Images d'une Somme d'Opérateurs Monotones et Applications. *Israël Journal of Mathematics*, Vol. 23, N°2, pp. 165-186, 1976.

[35] F.E. BROWDER. Nonlinear Operators and Nonlinear Equations of Evolution in Banach Spaces. *Proceedings of Symposia in Pure Mathematics, Vol. XVIII, Part 2, American Mathematical Society*, 1976.

[36] F.E. BROWDER and P. HESS. Nonlinear Mappings of Monotone Type in Banach Spaces. *Journal of Functional Analysis*, 11, pp. 251-294, 1972.

[37] G. BUTTAZZO. Semicontinuity, Relaxation and Integral Representation in the Calculus of Variations. *Pitman Research Notes in Mathematics*, Series, N° 207, 1989.

[38] G. BUTTAZZO and F. TOMARELLI. Compatibility Conditions for Nonlinear Neumann Problems. *Advances in Mathematics*, Vol. 89, N°2, pp. 127-143, 1991.

[39] B.D. CALVERT and C.P. GUPTA. Nonlinear Elliptic Boundary Value Problems in L^p-Spaces and Sums of Ranges of Accretive Operators. *Nonlinear Analysis, Theory, Methods & Applications*, Vol. 2, N°1, pp. 1-26, 1978.

[40] S. CARL. A Combined Variational-Monotone Iterative Method for Elliptic Boundary Value Problems with Discontinuous Nonlinearity. *Applicable Analysis*, Vol. 43, pp. 21-45, 1992.

[41] K.-C. CHANG. On the Multiple Solutions of the Elliptic Diffrential Equations with Discontinuous Nonlinear Terms. *Scienta Sinica*, Vol. XXI(N°2), 1976.

[42] K.-C. CHANG. Variational Methods for Non-Differentiable Functionals and Their Applications to Partial Differential Equations. *Journal of Mathematical Analysis and Applications*, 80, pp. 102-129, 1981.

[43] P.G. CIARLET. Quelques Remarques sur les Problèmes d'Existence en Elasticité Non Linéaire. *Rapport de Recherche INRIA*, N°121, 1982.

[44] P.G. CIARLET and J. NECAS. Unilateral Problems in Nonlinear Three-Dimensional Elasticity. *Arch. Rational Mech. Anal.*, 87, pp. 319-338, 1985.

[45] P.G. CIARLET and J. NECAS. Injectivity and Self-Contact in Nonlinear Elasticity. *Arch. Rational Mech. Anal.*, 97, pp. 171-188, 1987.

[46] P.G. CIARLET and P. RABIER. Les équations de von Kármán. *Lectures Notes in Mathematics 836, Springer Verlag*, 1980.

[47] A. CIMETIERE. Un problème de Flambement Unilatéral en Théorie des Plaques. *Journal de Mécanique*, Vol. 19, N°1, pp. 183-202, 1980.

[48] F.H. CLARKE. Nonsmooth Analysis and Optimization. *Wiley, New York*, 1984.

[49] M.G. CRANDALL and P.L. LIONS. Viscosity Solutions of Hamilton-Jacobi Equations. *Trans. Amer. Math. Society*, 277, pp. 1-42, 1983.

[50] B. DACOROGNA. Weak Continuity and Weak Lower Semicontinuity of Nonlinear Functionals. *Springer-Verlag, Berlin and New York*, 1982.

[51] R. DAUTRAY and J.-L. LIONS. Analyse Mathématique et Calcul Numérique pour les sciences et les techniques. *Masson Paris*, 1988.

[52] J.-P. DEDIEU. Cône Asymptote d'un Ensemble Non Convexe. Application à l'Optimisation. *C.R. Acad. Sci. Paris*, 287, pp. 91-103, 1977.

[53] J. DIEUDONNE. Eléments d'Analyse. *Gauthier-Villars*, 1968.

[54] P. DRABEK. Landesman-Lazer Type Condition and Nonlinearities with Linear Growth. *Czechoslovak Mathematical Journal*, 40, pp. 70-86, 1988.

[55] G. DUVAUT and J.L. LIONS. Les Inéquations en Mécanique et en Physique. *Dunod, Paris*, 1972.

[56] I. EKELAND and R. TEMAM. Convex Analysis and Variational Problems, *Studies in Mathematics and its Applications. American Elsevier Publishing Company, INC.-New York*, 1976.

[57] L.C. EVANS. Weak Convergence Methods for Nonlinear Partial Differential Equations. *Conference Board of the Mathematical Sciences, Regional Conference Series in Mathematics*, N°74, 1990.

[58] L.C. EVANS and R.F. GARIEPY. Measure Theory and Fine Properties of Functions. *Studies in Advanced Mathematics*, CRC Press, 1992.

[59] G. FICHERA. Problemi Elastostatici con Vincoli Unilaterali: il Problema di Signorini con Ambigue Condizioni al Contorno. *Mem. Acad. Naz. Lincei*, VIII 7, pp. 91-114, 1964.

[60] G. FICHERA. The Signorini Elastostatics Problem with Ambiguous Boundary Conditions. *In: Proc. Int. Conf. on the Application of the Theory of Functions in Continuum Mechanics*, Vol. I, Tbilisi, 1963.

[61] G. FICHERA. Boundary Value Problems in Elasticity with Unilateral Constraints. *Handbuch der Physik, VIa.2, Springer-Verlag, Berlin Heidelberg New York*, pp. 347-389, 1972.

[62] I. FONSECA. Variational Methods for Elastic Crystals. *Arch. Rational Mech. Anal.*, 97, pp. 189-220, 1987.

[63] S. FUCIK. Solvability of Nonlinear Equations and Boundary Value Problems. *Reidel, Dordrecht*, 1980.

[64] F. GASTALDI and F. TOMARELLI. Some Remarks on Nonlinear and Noncoercive Variational Inequalities. *Bollettino U.M.I.*, (7), 1-B, pp. 143-165, 1987.

[65] M. GIAQUINTA and E. GIUSTI. Researches on the Equilibrium of Masonry Structures. *Arch. Rational Mech. Anal.*, 88, pp. 359-392, 1985.

[66] E. GINER. Etudes sur les Fonctionnelles Intégrales. *Thèse de l'Université de Pau et des Pays de l'Adour*, 1985.

[67] D. GOELEVEN. On the Solvability of Linear Noncoercive Variational Inequality in Separable Hilbert Spaces. *Journal of Optimization Theory and Applications*, Vol. 79, N°3, pp. 493-511, 1993.

[68] D. GOELEVEN. On Noncoercive Variational Inequalities and some Applications in Unilateral Mechanics. *Ph. D. Degree in Science, F.U.N.D.P., Belgium*, 1993.

[69] D. GOELEVEN. On the Hemivariational Inequality Approach to Nonconvex Constrained Problems in the Theory of von Kármán Plates. *Z.A.M.M.*, Vol. 75, N°11, pp. 861-866, 1995.

[70] D. GOELEVEN. A Bifurcation Theory for Nonconvex Unilateral Plate Problem Formulated as an Hemivariational Inequality Involving a Potential Operator. *To appear in Z.A.M.M.*

[71] D. GOELEVEN. Noncoercive Hemivariational Inequality Approach to Constrained Problems for Star-Shaped Admissible Sets. *To appear in Journal of Global Optimization.*

[72] D. GOELEVEN. Noncoercive Hemivariational Inequalities and its Applications in Nonconvex Unilateral Mechanics. *To appear in Applications of Mathematics.*

[73] D. GOELEVEN, V.H. NGUYEN and M. WILLEM. Existence and Multiplicity Results for Semicoercive Unilateral Problems. *Bull. of the Australian Math. Society,* 49, pp. 489-498, 1994.

[74] D. GOELEVEN, V.H. NGUYEN and M. THERA. Méthodes du Degré et Branches de Bifurcation dans des Inéquations du type de Von Kármán. *Comptes Rendus de l' Académie des Sciences de Paris,* t. 317, Série I, p. 631-635, 1993.

[75] D. GOELEVEN, V.H. NGUYEN and M. THERA. Nonlinear Eigenvalue Problem Governed by Variational Inequality of Von Kármán Type: a Degree Theoretic Approach. *Topological Methods in Nonlinear Analysis,* Vol. 2, N°2, pp. 253-276, 1993.

[76] D. GOELEVEN and P.D. PANAGIOTOPOULOS. On a class of noncoercive hemivariational inequalities arizing in nonlinear elasticity. *F.U.N.D.P. Report,* 1995.

[77] D. GOELEVEN, G.E. STAVROULAKIS and P.D. PANAGIOTOPOULOS. Solvability Theory for a Class of Hemivariational Inequalities Involving Copositive Plus Matrics, Applications in Robotics. *To appear in Mathematical Programming.*

[78] D. GOELEVEN, D. MOTREANU and P.D. PANAGIOTOPOULOS. Multiple Solutions for a Class of Eigenvalue Problems in Hemivariational Inequalities. *To appear in Nonlinear Analysis, T.M.A.*

[79] D. GOELEVEN and M. THERA. Semicoercive Variational hemivariational Inequalities. *Journal of Global Optimization*, 6, pp. 367-381, 1995.

[80] A. HARAUX. Opérateurs Maximaux Monotones et Oscillations Forcées Non Linéaires. *Thèse de Doctorat d'Etat, Paris 6*, 1978.

[81] B. HERON and M. SERMANGE. Non Convex Methods for Computing Free Boundary Equilibria of Axially Symmetric Plasmas. *Rapport de Recherche INRIA*, N°108, 1981.

[82] P. HESS. On semi-coercive Nonlinear Problems. *Indiana University Mathematical Journal*, Vol. 23, N°7, pp. 645-654, 1974.

[83] H.N. KARAMANLIS and P.D. PANAGIOTOPOULOS. The Eigenvalue Problems in Hemivariational Inequalities and its Applications to Composite Plates. *Journal Mech. Behaviour of Materials*, vol. 3, pp.151-175, 1993.

[84] N. KIKUCHI and J.T. ODEN. Contact Problems in Elasticity: A Study of Variational Inequalities and Finite Element Methods. *SIAM, Philadelphia*, 1988.

[85] D. KINDERLEHER and G. STAMPACCHIA. An Introduction to Variational inequalities. *Academic Press, New York*, 1980.

[86] S.J. KIM and J.T. ODEN. Generalized Potentials in Finite Elastoplasticity. *Int. J. Engng. Sci.*, Vol. 22, N°11/12, pp. 1235-1257, 1984.

[87] S.J. KIM and J.T. ODEN. Generalized Flow Potentials in Finite Elastoplasticity-II. *Int. J. Engng. Sci.*, Vol. 23, N°5, pp. 515-530, 1985.

[88] R.J. KNOPS. Nonlinear Analysis and Mechanics. *Heriot-Watt Symposium*, Vol. 1, Pitman, 1977.

[89] R.J. KNOPS and C.A. STUART. Quasiconvexity and Uniqueness of Equilibrium Solutions in Nonlinear Elasticity. *Arch. Rational Mech. Anal.*, 86, pp. 233-249, 1986.

[90] M.S. KUCZMA and E. STEIN. On Nonconvex Problems in the Theory of Plasticity. *Arch. Mech.*, 46, 4, pp. 505-529, 1994.

[91] E.M. LANDESMAN and A.C. LAZER. Nonlinear Perturbations of Linear Elliptic Boundary Value Problems at Resonance. *Journal of Mathematics and Mechanics*, 19, pp. 609-623, 1970.

[92] P.L. LIONS. Two Remarks on the Convergence of Convex Functions and Monotone Operators. *Nonlinear Analysis, Theory, Methods & Applications*, Vol. 2, N°5, pp. 553-562, 1978.

[93] J.L. LIONS and G. STAMPACCHIA. Variational Inequalities. *Comm. Pure Applied Math.*, XX, pp. 493-519, 1967.

[94] D.T. LUC. Recession Cones and the Domination Property in Vector Optimization. *Mathematical Programming*, 49, pp. 113-122, 1990.

[95] D.T. LUC. Recession Maps and Applications. *Optimization*, 27, pp. 1-15, 1993.

[96] D.T. LUC and J-P. PENOT. Convergence of Asymptotic Directions. *Preprint Université de Pau*, 1994.

[97] J. MAWHIN. Points Fixes, Points Critiques et Problèmes aux Limites. *Sém. de Math. Sup., Les Presses de l'Université de Montréal*, 1985.

[98] J. MAWHIN. Semi-coercive Monotone Variational Problems. *Bulletin de la Classe des Sciences*, 5e série - Tome LXXIII, 3-4, pp. 118-130, 1987.

[99] J. MAWHIN. Analyse, Fondements, Techniques, Evolutions. *De Boeck Université*, 1992.

[100] J. MAWHIN and K. SCHMITT. Landesman-Lazer Type Problems at an Eigenvalue of Odd Multiplicity. *Results in Math.*, 14, pp. 138-146, 1988.

[101] J. MAWHIN and M. WILLEM. Critical Point Theory and Hamiltonian Systems. *Spinger Verlag, New York*, 1989.

[102] J.J. MOREAU. La Notion de Sur-potentiel et les Liaisons Unilatérales en Elastostatiques. *C.R. Acad. Sci. Paris*, Série A, 267, pp. 954-957, 1968.

[103] D. MOTREANU and Z. NANIEWICZ. Discontinuous Semilinear Problems in Vector Valued Function Spaces. *To appear in Diff. and Integral Equations.*

[104] Z. NANIEWICZ. On Some Nonmonotone Subdifferential Boundary Conditions in Elastostatics. *Ingenieur-Archiv.*, 60, pp. 31-40, 1989.

[105] Z. NANIEWICZ. On the Pseudomonotonicity of Generalized Gradients of Nonconvex Functions. *Applicable Analysis*, Vol. 47, pp. 151-172, 1992.

[106] Z. NANIEWICZ. Hemivariational Inequality Approach to Constrained Problems for Admissible Sets. *Journal of Optimization, Theory and Applications*, Vol. 83, N°1, pp. 97-112, 1994.

[107] Z. NANIEWICZ. Hemivariational Inequalities with Functions Fulfilling Directional Growth Condition. *Applicable Analysis, 55, 3-4, pp. 259-285*, 1994.

[108] Z. NANIEWICZ and P.D. PANAGIOTOPOULOS. The mathematical Theory of Hemivariational Inequalities and Applications. *Marcel Dekker, N. York*, 1994.

[109] J. NECAS. Méthodes Directes en Théorie des Equations Elliptiques. *Masson, Paris*, 1967.

[110] P.D. PANAGIOTOPOULOS. Nonconvex Superpotentials in the Sense of F.H. Clarke and Applications. *Mech. Res. Comm.*, 8, pp. 335-340, 1981.

[111] P.D. PANAGIOTOPOULOS. Noncoercive Energy Function, Hemivariational Inequalities and Substationarity Principles. *Acta Mech.*, 48, pp. 160-183, 1983.

[112] P.D. PANAGIOTOPOULOS. Inequality Problems in Mechanics and Applications, Convex and Nonconvex Energy Functions. *Birkhaüser, Basel*, 1985.

[113] P.D. PANAGIOTOPOULOS. Nonconvex Problems of Semipermeable Media and Related Topics. *ZAMM*, 65, 1, pp. 29-36, 1985.

[114] P.D. PANAGIOTOPOULOS. Hemivariational Inequalities and Substationarity in the Static Theory of v. Kármán Plates. *ZAMM*, 65, 6, pp. 219-229, 1985.

[115] P.D. PANAGIOTOPOULOS. Hemivariational Inequalities and their Applications, In: J.M. Moreau, P.D. Panagiotopoulos, G. Strang (eds), Topics in Nonsmooth Mechanics. *Birkhaüser Verlag*, 1988.

[116] P.D. PANAGIOTOPOULOS. Inéquations Hémivariationnelles Semi-coercives dans la Théorie des Plaques de Von Kármán. *C.R. Acad. Sci. Paris*, t. 307, Série I, p. 735-738, 1988.

[117] P.D. PANAGIOTOPOULOS. Semicoercive Hemivariational Inequalities, On the Delamination of Composite Plates. *Quaterly of Applied Mathematics*, Vol. XLVII, N°4, pp. 611-629, 1989.

[118] P.D. PANAGIOTOPOULOS. Coercive and Semicoercive Hemivariational Inequalities. *Nonlinear Analysis, Theory, Methods & Applications*, Vol. 16, N° 3, pp. 209-231, 1991.

[119] P.D. PANAGIOTOPOULOS. Hemivariational Inequalities, Applications in Mechanics and Engineering. *Springer Verlag, Berlin, Heidelberg*, 1993.

[120] P.D. PANAGIOTOPOULOS and G.E. STRAVOULAKIS. A Variational-Hemivariational Inequality Approach to the Laminated Plate Theory under Sub-differential Boundary Conditions. *Quaterly of Applied Mathematics*, XLVI, 3, pp. 409-430, 1988.

[121] P.D. PANAGIOTOPOULOS and G.E. STRAVOULAKIS. The Delamination Effect in Laminated Von Kármán Plates Under Unilateral Boundary Conditions. A Variational-Hemivariational Inequality Approach. *Journal of Elasticity*, 23, pp. 69-96, 1990.

[122] D. PASCALI and S. SBURLAN. Nonlinear Mappings of Monotone Type. *Sijthoff and Noordhoff International Publishers, Amsterdam, The Netherlands*, 1978.

[123] B.D. REDDY and F. TOMARELLI. The Obstacle Problem for an Elastoplastic Body. *Appl. Math. Optim.*, 21, pp. 89-110, 1990.

[124] R.T. ROCKAFELLAR. Convex Analysis. *Princeton University Press, Princeton, NJ*, 1970.

[125] M. SCHATZMAN. Problèmes aux Limites Non Linéaires, Non Coercifs. *Ann. Scuola Norm. Sup. Pisa, Cl. Sci.*, 27, pp. 641-686, 1973.

[126] P. SHI and M. SHILLOR. Noncoercive Variational Inequalities with Applications to Friction Problems. *Proceedings of the Royal Society of Edinburgh*, vol. 117A, pp. 275-293, 1991.

[127] G. STAMPACCHIA. Equations Elliptiques du Second Ordre à Coefficients Discontinus. *Sém. de Math. Sup. Les Presses de l'Université de Montréal*, 1966.

[128] G.E. STAVROULAKIS, D. GOELEVEN and P.D. PANAGIOTOPOULOS. New Models for a Class of Adhesive Grippers. The Hemivariational Inequality Approach. *To appear in Ingenieur Archiv.*, 1995.

[129] A. SZULKIN. Minimax Principles for Lower Semicontinuous Functions and Applications to Nonlinear Boundary Value Problems. *Ann. Inst. Henri Poincaré*, Vol 3, N°2, pp. 77-109, 1986.

[130] R. TEMAM. Mathematical Problems in Plasticity. *Gauthier Villars, Paris*, 1985.

[131] F. TOMARELLI. A Quasi-Variational Problem in Nonlinear Elasticity. *Annali di Matematica Pura et Applicata*, (IV), Vol. CLVIII, pp. 331-389, 1991.

[132] F. TOMARELLI. Noncoercive Variational Inequalities for Pseudomonotone Operators. *Rend. Sem. Mat. Fis. Univ. Milano*, n.83/P, 1993.

[133] A. TYCHONOV. Solution of incorrectly formulated problems and the regularization method. *Soviet Math. Dokl.*, 4, pp. 1035-1038, 1963.

[134] M. M. VAINBERG. Variational Method and Method of Monotone Operators in the Theory of Nonlinear Equations. *John Wiley & Sons, New York*, 1973.

[135] L.K. VY and K. SCHMITT. Minimization Problems for Noncoercive Functionals Subject to Constraints. *To appear in Trans. Amer. Math. Soc.*

[136] M. WILLEM. Analyse Convexe et Optimisation. *Cabay, Louvain-la-Neuve*, 1986.

[137] Cz. WOŹNIAK. Materials with Generalized Constraints. *Arch. Mech.*, 36, 4, pp. 539-551, 1984.

[138] C. ZALINESCU. Recession Cones and Asymptotically Compact Sets. *J. Optim. Theory Appl.*, 77, pp. 209-220, 1993.

[139] E. ZEIDLER. Nonlinear Functional Analysis and its Applications III. *Variational Methods and Optimization, New York*, 1984.

[140] K. ZHANG. Energy Minimizers in Nonlinear Elastostatics and the Implicit Function Theorem. *Arch. Rational Mech. Anal.*, 114, pp. 95-117, 1991.

BIBLIOGRAPHICAL Remark: A substantial part of each of the following books concerns the study of semicoercive unilateral problems..

- G. **Duvaut and J.L. Lions** , *Les Inéquations en Mécanique et en Physique,* Dunod, Paris 1972.

- I. **Hlavacek, J. Haslinger, J. Nečas and J. Lovisek** , *Solutions of Variational Inequalities in Mechanics* , Applied Mathematical Sciences 66, Springer Verlag, New York 1982.

- N. **Kikuchi and J.T. Oden** , *Contact Problems in Elasticity: A Study of Variational Inequalities and Finite Element Methods* , SIAM, Philadelphia 1988.

- Z. **Naniewicz and P.D. Panagiotopoulos** , *The mathematical Theory of Hemivariational Inequalities and Applications* , Marcel Dekker, N. York 1994.

- P.D. **Panagiotopoulos,** *Hemivariational Inequalities, Applications in Mechanics and Engineering* , Springer Verlag Berlin Heidelberg 1993.

- P.D. **Panagiotopoulos** , *Inequality Problems in Mechanics and Applications, Convex and Nonconvex Energy Functions* , Birkhäuser, Basel 1985.

- J.F. **Rodriguez** , *Obstacle Problems in Mathematical Physics* , Mathematics Studies n°134, Elsevier Science Publishers B.V., 1987.

- G.M. **Troianiello** , *Elliptic Differential Equations and Obstacle Problems* , Plenum Press, New York 1987

6   NOTATIONS

Latin indices take their values in the set {1,2,3} and the repeated index convention is used.

$\cdot$	scalar product in $I\!\!R^N$
$\wedge$	vector product in $I\!\!R^N$
$u = (u_i)$	vector with components u_i
$A = (A_{ij})$	matrix with elements A_{ij}
$A : B := A_{ij}B_{ij}$	matrix inner product
$\mathrm{adj}(A)$	adjugate of the matrix A, i.e. the transpose of the cofactor matrix
$\det(A)$	determinant of A
M^3	set of all matrices of order 3
$\nabla u = (v_{,i}) = \mathrm{grad}(v)$	gradient of a mapping $v : I\!\!R^3 \to I\!\!R$
$\nabla E = (E_{i,j})$	gradient of a mapping $E : I\!\!R^3 \to I\!\!R^3$
$\mathrm{div}(T) = (T_{ij,j})$	divergence of a tensor field $T : I\!\!R^3 \to M^3$
$D^2 F$	Hessian of a mapping $F : I\!\!R^N \to I\!\!R$
$\partial u / \partial n$	normal derivative of $u : I\!\!R^N \to I\!\!R$
$\dot{u}$	derivative of $u : I\!\!R \to I\!\!R$
∂f	convex subdifferential if f is proper convex and l.s.c.
∂f	Clarke's subdifferential if f is locally Lipschitz
$\frac{\partial W}{\partial F}(F)$	gradient of a mapping $W : M^3 \to I\!\!R$
Ω	open bounded set in $I\!\!R^N (N \geq 1)$
Γ	boundary of Ω

$\mu(\Omega)$	N-dimensional Lebesgue measure
H_N	N-dimensional Hausdorff measure
$\cdot dx$	volume element in $I\!\!R^N(\mu)$
ds	area element in $I\!\!R^N(H_{N-1})$
$\|\cdot\|_{1,p}$	norm in the space $W^{1,p}(\Omega)$ or $W^{1,p}(\Omega; I\!\!R^n)$
$\|\cdot\|_{0,p}$	norm in the space $L^p(\Omega)$ or $L^p(\Omega; I\!\!R^n)$
$\|\cdot\|_{\Gamma,0,p}$	norm in the space $L^p(\Gamma)$ or $L^p(\Gamma; I\!\!R^n)$
$X \hookrightarrow Y$	the identity map i from X to Y is continuous and $X \subset Y$
$X^\perp$	orthogonal space of X
$\langle X, X' \rangle$	duality product between a real Banach space and its topological dual X', scalar product when X is a real Hilbert space $(X = X')$
(σ_{ij})	stress tensor
$\varepsilon(u) = (\varepsilon_{ij}(u))$	linearized strain tensor whose components are given by $\varepsilon_{ij}(u) = \frac{1}{2}\{u_{i,j} + u_{j,i}\}$.
dom$\{f\}$	domain of a function $f : X \to I\!\!R \cup \{+\infty\}$ $= \{x \in X \mid f(x) < +\infty\}$
$Ker\{f\} = f^{-1}(0)$	kernel of a function $f : X \to I\!\!R \cup \{+\infty\}$ $= \{x \in X \mid f(x) = 0\}$
$f(x+0)$	limit of $f(x)$ as $x \to 0^+$
$f(x-0)$	limit of $f(x)$ as $x \to 0^-$
supp $\{f\}$	support of f
Ker(A)	kernel of an operator $A : X \to X'$
$R(A)$	range of an operator $A : X \to 2^{X'}$
graph(A)	graph of an operator $A : X \to 2^{X'}$

$\overline{D}$	closure of a set D
$\text{int}\{D\}$	interior of a set D
$D^c = X \backslash D$	complementary of a set $D \subset X$
$f * g$	convolution of f and g
$\Delta(\varepsilon_n)$	see p. 22
$R(\Delta(\varepsilon_n))$	recession set, p. 23
D_μ	see p. 24
$a(\tau)$-compact	asymptotically compact with respect to the topology τ, p. 24

T - #0238 - 071024 - C0 - 248/170/10 - PB - 9780582304024 - Gloss Lamination